T0357291

The Amazing Beginnings
of Mathematics

A Mathema Publishing
Trust Publication

—

The Amazing Beginnings of Mathematics

mathema
PUBLISHING TRUST

Hardie Grant
BOOKS

Preface

This book is the first production of the Mathema Publishing Trust, an entity that has been established to provide support for the Mathema Foundation. The Foundation is an educational charity that has two main purposes:

- to remove much of the mystery associated with mathematics and the misperception that mathematics is only symbolic and difficult
- to inform schools and the wider community about methods that will ensure more individuals understand and appreciate this important discipline.

The first project of the Foundation has been to establish the Mathema Gallery, a physical venue where members of the community can visually and interactively learn about the peoples, places, events and developments that have influenced the mathematics we use today. The Gallery will house one of the largest and most extensive private mathematics libraries in Australia. This includes over 800 titles published before 1850.

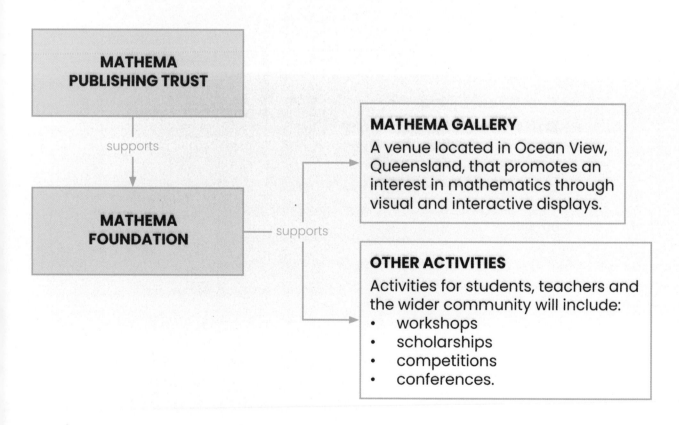

MATHEMA PUBLISHING TRUST

supports

MATHEMA FOUNDATION

supports

MATHEMA GALLERY
A venue located in Ocean View, Queensland, that promotes an interest in mathematics through visual and interactive displays.

OTHER ACTIVITIES
Activities for students, teachers and the wider community will include:
- workshops
- scholarships
- competitions
- conferences.

Mathematics has an interesting and rich early history that evolved in different ways in at least six locations: Central America, South America, Africa, the Middle East, the Far East and South Asia. People living in these separate regions had some common concerns that required the need for similar mathematical ideas, such as the need to count and record the cycles of events that allowed them to plant and harvest crops. As a result, each culture developed a **number system** for use in daily affairs. Interestingly, and in complete isolation, the separately developed number systems had features that were the same. Number groupings were commonly in tens, hundreds, thousands and so on. **Place value** systems for numbers existed for number systems that used **base 10** (in the Indus Valley), **base 60** (Mesopotamia) and **base 20** (Maya). The invention of a zero was a challenge.

The oldest of the number systems was based on groupings of 60. The reasons for this choice seem to have been purely mathematical: to allow greater flexibility to work with fractional amounts. While this choice might be questioned, it heavily influenced the mathematics we use today. **Time** is recorded in blocks of 60 **minutes** and 60 **seconds**. **Angles** are measured in **degrees** with six groups of 60 making one complete revolution. This mathematics was established over 4,000 years ago and has remained in use ever since.

Construction of buildings such as **pyramids** was common in at least four of the six early cultures. Mathematics enabled these to be built accurately, and structures were often oriented in specific directions. As a result, the **triangle** shape became an important tool. The **right-angled triangle** was the most important of these shapes. Facts were recorded and techniques developed for using this triangle long before the Greek civilisation existed, even though we now talk about the Pythagorean triangle or theorem. It could just have well been called the Tigris triangle (from the Tigris River) or the Nile triangle or the Yangtze triangle. Right-angle triangle mathematics, which we call **trigonometry**, is based on these 3,000-year-old facts.

Many mathematical applications today relate to digital technologies. At its basic level, this involves work in a **base-2** (**binary**) system. Over 2,000 years ago, this system was used in the Indus Valley as a system of weights. The Hindus showed that binary was more efficient than the other system (base-10) they were using for numbers. This is still true today.

Many other modern mathematics topics can be connected to the six early civilisations that are described in this book. The book has been designed to involve the reader as topics are explored, and the combination of text and activities presents how the mathematics was used "then". We hope this gives readers an appreciation of the sophisticated level of the mathematics that predates, and is often erroneously credited to, peoples who lived in Europe.

Introduction

When? Where? How?
Who? Why?
I wonder, wonder,
wonder, wonder.

What branches of mathematics are truly important today?

What geometry does a global positioning system (GPS) use in a three-dimensional world?

Why do modern mathematics curricula, even in the earliest grades, include algebra?

What is special about our measurement system?

What is data science and why is it important?

What mathematics did early civilisations develop to ensure their longevity and stability?

What methods did early civilisations use to communicate numerical information?

The mathematics we use today

How often do you think you use mathematics in your everyday life? You may think you are using mathematics less now that we have so much technology in our lives, but actually we use it so much more!

Your day may begin with you looking at a clock that shows numbers around the edge of a circle or on a digital display. Looking outside, you may judge the temperature to be a certain number of degrees, and the few wispy clouds in the sky mean there is a low chance of rain. You only have time for 20 minutes of exercise, so the morning run is restricted to just 3 kilometres (km). Your wrist monitor, which is linked to a GPS, keeps you informed of your heart rate, speed, distance and a whole host of other things. Later, you look online to check different news and social media sites. Your logins require various numerical user identifications and passwords, which are encrypted when sent, then decrypted by the recipient. You make a video call to a friend who lives in another part of the world, but because your friend lives

in a different time zone it is now very late at night, so you don't talk for long. The day has just begun, but you have already used many different aspects of mathematics. What else lies ahead?

It needs to be stated here: we all underestimate the importance of mathematics. It is universally agreed that we need a basic understanding of numbers, operations, measurement, geometry and statistics for use in everyday situations. Mathematics has been, and always will be, the enabler of the way we live today, and the extent of our mathematical knowledge determines our capacity to make decisions and choices beyond the basic requirements.

Numbers and number systems

Numbers are the underlying DNA of mathematics. Our current number system is used nearly universally. Different countries and cultures have different words for the numbers and some languages have a set of symbols for numbers. But the Hindu–Arabic numerals are recognised everywhere and are the basic language of mathematics.

The Hindu–Arabic system became popular for two reasons. First, the system uses just 10 symbols (digits) so it is easy to learn and remember the values. The second reason is the place-value structure of the system – the value of the digit depends on the place where it is written. The value of the places has a predictable growing pattern of increasing powers of 10 from right to left: 10, 10 × 10, 10 × 10 × 10, 10 × 10 × 10 × 10 and so on. Hence the Hindu–Arabic numbers can be described as a base-10 place-value system.

Introduction

This 4 has a value of 4 × 10 = 40

$$4,246$$

This 4 has a value of
4 × (10 × 10 × 10) = 4,000

Whole numbers were the first developed and are the foundation of mathematics. They are the numbers used to describe how many items are in a collection of objects: zero, one, two, three, four and so on.

There are five llamas in my herd.

The most powerful feature of the number system we use today is that it provides a logical method to write a number for any size collection of objects that can be counted. There is no end to the set of whole numbers!

Over time, it has been necessary to include new types of numbers. For example, financial transactions meant that balances could be less than zero, so it was necessary to introduce negative values. Collectively, the positive and negative whole numbers are called integers. Ancient mathematicians had the idea of fractions (known as rational numbers) but found them difficult to explain or express.

Irrational numbers such as π (pi) (which cannot be written as a fraction of two integers) were next to impossible to understand even though the Greeks knew they had to exist. The rational and irrational are collectively called real numbers. In this combined collection, decimals are used to show both types of numbers. In fact, real numbers are essentially every possible decimal that can ever be written – some stop, some go on forever in a repeating pattern and some go on forever and never repeat. (Now that's a lot of numbers!) There are also other numbers, which could be described loosely as unreal or complex.

In this book, we discuss what motivated and helped people from different cultures to develop all these numbers and other strands of mathematics. We start by looking at the five main strands of mathematics:

- operations with numbers (arithmetic)
- geometry
- measurement
- statistics and probability
- algebra.

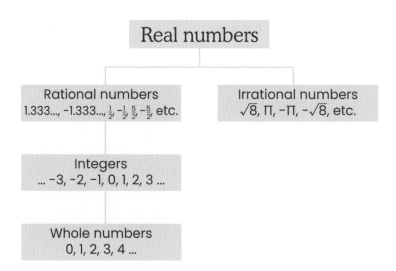

Introduction

Operations with numbers (arithmetic)

The word *mathematics* usually conjures images that involve computation – addition, subtraction, multiplication or division. There are other operations in mathematics, but these are the first that students learn. Today, many machines (such as calculators) can complete these four basic tasks, so the need to master these and even more sophisticated computational skills is waning.

The computational methods that have been taught in schools over the past 200 years depend on the structure of the number system and took a long time for the formal procedures to become established. Before the current number system, which enabled written algorithms that could be performed by paper and pencil, calculations were completed using tools such as an abacus. This device was necessary because the Roman numerals that were the dominant symbolic number system at the time were too cumbersome. The 16th century image on the next page shows a competition between an algorist (using a pen to write numerals) and an abacist (who uses counters on a board or an abacus) to see which was quicker. Algorithms with the modern numbers won! Today, we are still devising methods to complete complex calculations quickly, but these modern-day algorithms are all constructed for use by digital devices and are very different from the algorithms of the past.

In the 21st century it isn't crucial that students learn all the paper and pencil methods to calculate that were emphasised in schools just two decades ago. It is important that students are proficient with some of the key mental methods to calculate. These methods do rely on knowing numbers and some of the many properties of the basic concepts of operations. For example, in a real world problem which might initially be written as 345 − 198 = ___ , it is often easier to rewrite the equation as 198 + ___ = 345. In this case, the answer can be found by counting on from the 'smaller' number, which is very often faster than reaching for a calculator. The same thinking can be applied to a division problem.

In summary, knowledge about operations is important today, but the focus has shifted from emphasising rote procedures to learning all the important conceptual aspects of an operation. Rote procedures can be completed by digital devices, but these electronic tools need to be programmed in efficient ways, often using key conceptual properties of operations.

Geometry

Geometry does not have the same level of recognition as numbers or operations in mathematics. However, in many ways geometry lets people classify objects in some way (such as by shape, colour or function), which allows them to be counted, added, subtracted etc. Geometry first involves the ideas of *same* and *different* in relation to everyday shapes. From this starting point, things can be classified more formally as two-dimensional (2D) shapes or three-dimensional (3D) objects. Numbers are really only needed to describe a collection of things that are the same in some way. The counted objects may be real world items, and they will nearly always involve some form of geometry. Hence, geometry is the first aspect of mathematics that people encounter.

The word *geometry* means earth (*geo*) measure (*metry*). One of the first applications of mathematics was indeed measuring the earth, so the word *geometry* is appropriate. Nearly all geometry taught until the 1950s was originally described by Euclid, a Greek mathematician who lived about 2,300 years ago. The rules for this work were based on a world that was flat and fixed. For this reason, early studies of geometry involved 2D shapes, which then later progressed to 3D objects. Two-dimensional shapes and 3D objects are linked because 3D objects are built from 2D shapes. For example, the picture below shows 3D prisms with the names of the 2D shapes (faces) that distinguish between and make up the prisms.

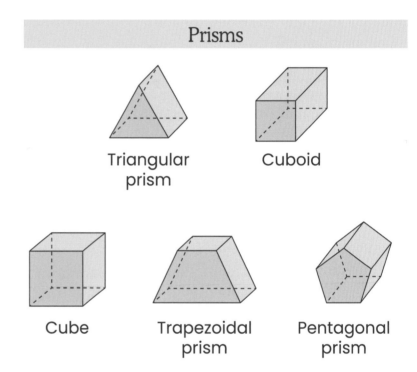

Prisms

Triangular prism

Cuboid

Cube

Trapezoidal prism

Pentagonal prism

Working with geometric shapes is relatively easy if you know the terminology and the connection between various shapes and objects. Polygons are usually described as 2D shapes with straight sides, although the word *polygon* translates directly as "many angles". The diagram at the top of the next page shows the different categories of quadrilaterals (polygons with four sides) and the connection between the various types. Polygons also include many other subcategories, such as triangles and pentagons.

Quadrilaterals

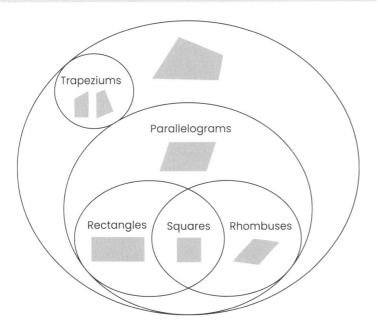

ACTIVITY 1:
Sorting and naming triangles

Do you know the names for all of the different triangles? Go to page 32 and complete the activity.

Polygons are a subcategory within a larger group of 2D shapes that includes shapes with sides that are not straight. Collectively, shapes with straight and/or curved sides are known as closed shapes.

Interestingly, one of the easiest shapes to draw is the circle. A stick or two and a piece of string are the only tools required to construct a perfect circle. Note that the line you have drawn is a 2D shape; the interior is not part of the shape.

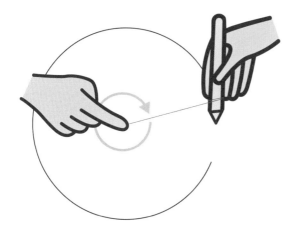

Once you have drawn a circle, you can draw all sorts of other 2D shapes. You can use the distance from the centre to the circle (the line) to mark six equal arcs around the edge. These points can be connected to make various shapes.

In the same way that numbers form a hierarchy from whole numbers to real numbers (see page 8), there is an overall hierarchy for geometry. Today, the study of geometry is divided into Euclidean and non-Euclidean geometry. Euclidean geometry was formalised by the Greeks, who worked with rules that stated that the sum of the angles in every triangle was 180 degrees. More recent analysis has found situations that vary from the work of the ancient Greeks. Non-Euclidean geometry was developed after it was shown that the earth is not flat and that some triangles' angles could add up to more than 180 degrees. Non-Euclidean geometry has three main sub-branches.

Spherical geometry, a special type of elliptical geometry, enables aeroplanes to navigate and GPSs to operate on the outer surface of a sphere (a globe). In this geometry, the sum of the angles of a triangle drawn on a sphere is greater than 180 degrees.

Hyperbolic geometry is the opposite of spherical geometry in that the sum of the angles of a triangle drawn on a hyperbolic surface is less than 180 degrees. Hyperbolic geometry is used to analyse the growth of plants and at least one animal (coral). The investigations into hyperbolic geometry have led to an interesting intersection of mathematics and art. In 1997, Latvian mathematician and artist Dr Daina Taimina discovered a way to model hyperbolic planes through the art of crochet. These models closely resemble the growth pattern of coral. This groundbreaking work shows the basic principles of hyperbolic geometry in a mathematically friendly and artistic manner (see below).

Topology involves bending and twisting to show that objects that look different can really be classified as the same. This includes intriguing topics such as graph theory, networks, knots and fractals with many applications that are only beginning to emerge. For example, fractals can be used to create intriguing designs but they also have important applications in digital technologies. This includes creating clearer screen images and better antennas, such as the Vicsek fractal (shown at right) that improves mobile phone reception.

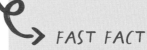

Measurement

The study of geometry quickly leads to the third strand of mathematics: measurement. It is difficult to discuss shapes without describing the various features that are evident. Numbers can be used to describe and compare the number of faces, edges and vertices of objects, but this discussion is limited unless there is a way to compare lengths of sides, areas of faces or volumes of interior spaces. Some features of objects such as faces are discrete (countable), so the number of faces is the measure. Lengths, areas, and spaces are continuous (measurements that are not countable), so they are difficult to compare unless a number is attached. Measurement involves assigning a number to an attribute or feature of a geometrical object or shape. Basic attributes such as length, weight, capacity or degrees are measured using a calibrated tool; for example, a ruler, scales, jug or compass. Some numbers that are assigned to other attributes such as area and volume are often derived using a rule or formula. The diagrams used on pages 14 and 15 to show the connection between shapes help guide the sequence in which students learn the various measurement rules.

Introduction

The boxes below show the connection between the key metric attributes, the sequence in which they are connected and the base unit for each attribute.

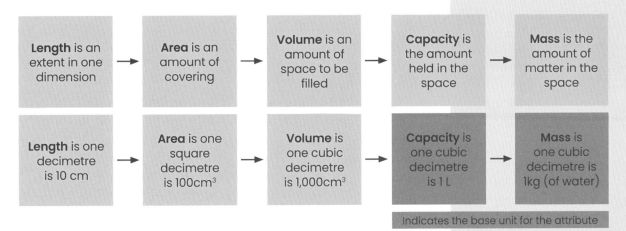

Length is an extent in one dimension	**Area** is an amount of covering	**Volume** is an amount of space to be filled	**Capacity** is the amount held in the space	**Mass** is the amount of matter in the space
Length is one decimetre is 10 cm	**Area** is one square decimetre is 100cm³	**Volume** is one cubic decimetre is 1,000cm³	**Capacity** is one cubic decimetre is 1 L	**Mass** is one cubic decimetre is 1kg (of water)

Indicates the base unit for the attribute

These diagrams show that there is a close connection between the measurement attributes and corresponding units. Prior to the 1800s, the measurement system was chaotic. The attributes were nearly the same everywhere, but the attributes did not relate to each other, so there was no connection between volume of a container, its capacity or its weight. A greater problem was that the standard units varied greatly from one country to the next and often between villages in the same region. For example, a foot wasn't the same length everywhere.

The metric system is an application of the decimal number system, which means conversion rules are easy to remember and apply across different attributes. The system began in France when Napoleon was emperor and over the next 200 years gradually spread to nearly all countries of the world. The system is mathematically logical and is used universally in science and medicine where there is a need for consistency and accuracy.

Usually, measurement is considered to be the assignment of a number to a thing – a part or a feature of a shape or

~ACTIVITY 2:
All about metric cubes

What do you know about this container and the amount of water it holds? Go to page 33 and complete the activity.

object. A number can be assigned to almost anything. A very common attribute is the amount of time assigned to an event. Today, it might be expressing the length of time it takes to drive or fly, watch a TV show or movie, run a certain distance, sleep at night and so on. Temperature is another attribute that has a number assigned to it and is a form of measurement. Early in the development of the metric system, there was an attempt to change time and angle measurements to a metric system. However, this proved to be too culturally difficult, so today angle measurement and time use the units that were established long ago. Of course, temperature is metricated.

In 1791 the French Academy of Sciences defined the metre as equal to one ten-millionth of the distance from the North Pole to the equator along the meridian through Paris. When the length of metre was decided, stone references like this example were installed throughout Paris so the population could learn about this standard measure. Note that this metre also includes decimetres.

Introduction

Statistics and probability

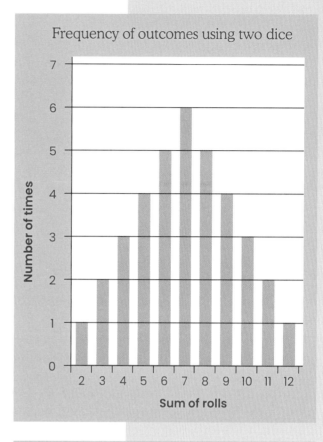

Frequency of outcomes using two dice

Number of times / Sum of rolls

M uch of statistics has been devised relatively recently, but the concept of probability was developed alongside the early work with numbers. Probability involves the likelihood of outcomes associated with certain random events such as tossing coins or rolling dice. Dice are interesting because the objects rolled could have a variable number of faces.

The situation could also involve rolling more than two dice. These graphs show how the outcomes change for two and three dice. When you analyse the data of these graphs, you are using statistics.

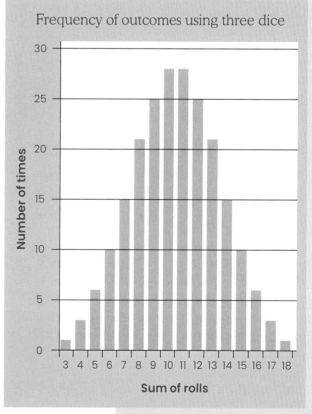

Frequency of outcomes using three dice

Number of times / Sum of rolls

Algebra

Algebra is usually associated with the mathematics of unknowns and originated about 1,500 years ago. The word *algebra* comes from part of the title of a work by a Middle-Eastern mathematician, Muhammad ibn Mūsā al-Khwārizmī. The Arabic term *al-jabr* is translated to mean "reunion of broken parts" and "to balance". These expressions might be interpreted as working with broken or unknown parts that are balanced (equal). This describes what is involved in the study of algebra. Note that al-Khwārizmī's name is the origin of the word *algorithm*.

The usual image of algebra involves equations with unknowns, usually x, y, n, p or q. However, the first work with algebra was predominantly words (symbols of a sort), which were used to discuss the methods. When the work of the early mathematics was shared with Europe through the writings of Leonardo of Pisa (also known as Fibonacci), the book was nearly all words.

The most important concept in the study of algebra is equality – ensuring that, in situations that were equal, things are still equal after an action has been completed. The ideas of algebra can be represented pictorially (as shown here) as well as in words, before they are written as abstract equations. This is a vital step that is too often hurriedly taught or totally skipped.

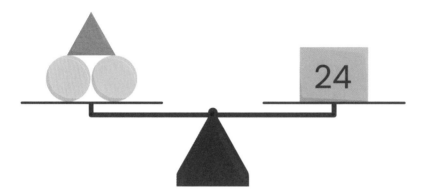

Over time, algebra has become very symbolic and, unfortunately, removed from the understanding and grasp of young learners. This means that it is even more important to spend time representing the ideas of algebra visually for young learners.

What, where and when did it all begin?

The development of mathematics has a fascinating history that is spread across four continents (Asia, Africa, North America and South America). The history is long, with recorded events spanning back over 5,000 years. Isolated artefacts, which are still being analysed by archaeologists, indicate some type of mathematics may even have occurred before this time. All of these developments occurred in parts of the world that today are off main travel routes or too dangerous to reach, so the details of many early contributions are still being identified. This book gives the reader up-to-date information on previously unknown mathematical facts. These new facts fit into the puzzle that is slowly being pieced together about the contributions older cultures made to the mathematics we use today. Some of the facts will surprise you!

The earliest civilisations developed in fertile regions where people could live without needing to move around to find food or water. They built communities where food-based commerce could occur. These communities had permanent structures that needed to be built and simple systems of trade that required records. As these regions became established, governments expanded to assume more responsibilities, which led to more roles that required more than simple record keeping. Buildings and public facilities needed to be constructed, systems of trade and exchange were established, and events such as seasons based on celestial observations were analysed. All of these early activities were the beginnings of mathematics.

The six key cultures discussed in this book worked with all strands of mathematics almost simultaneously, beginning more than 4,000 years ago. The map below shows the locations of the first true civilisations. Each had some degree of a formal governing structure that included a level of mathematical development. All of these cultures invented ways to record numbers, use geometry and apply mathematics to measure and in some cases to calculate.

These civilisations built roads, water systems, pyramids and temples. Their records told of the number of people living in various regions, the amount of grain that had been harvested, the animals owned by farmers and the dimensions of farmland. They also recorded the timing of events such the lengths of days, cycles of the moon or the timing of a solstice or equinox, which helped them plan celebrations. It took many centuries to devise the systems that were used, but this work was the setting for mathematics that is used today.

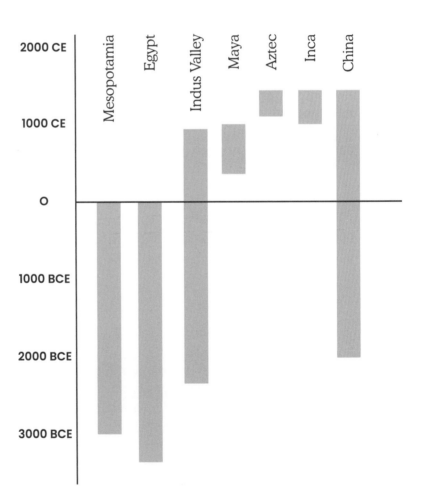

Note: The Maya, Aztec and Inca civilisations were preceded in their regions with groups that had started the development of mathematical ideas.

Number systems and calculations

Four thousand years ago, the civilisations developed their mathematical ideas in isolation, but in some ways the number systems they each developed were similar. The people in each region seemed to understand the idea of numbers and the need to numerate (count collections). Five of the regions shown in the map on page 23 developed a set of symbols used to express numbers. Of course, the symbols were different and the method for using the notation was also slightly different. People in the sixth region – the Andes mountains of South America – used objects rather than symbols to record numbers.

Many of the features of the number systems and a few other mathematical ideas that were created long ago in the regions are still with us today. How the individual regions developed and created their own way of expressing mathematics and how we use it today is a fascinating story. Of particular interest is how the people of the Indus Valley region developed their number system, which then gradually evolved over the next 3,000 years to become the system of numbers we use today.

Several things influenced the first number systems. One influence was the anatomy of the human body. The fact that we have two hands with a total of 10 fingers certainly influenced most of the early cultures. The other important consideration for a number system was the ability to divide it (the base of the system) into equal parts so there wasn't an amount left over. Dealing with an amount left over required fractions, which were difficult at that time to understand and challenging to write.

One number that divides into a second number without anything left over is called a *factor* of that second number. Ten has four factors: 1, 2, 5 and 10. Twelve has six factors: 1, 2, 3, 4, 6 and 12. Twenty also has six factors: 1, 2, 4, 5, 10 and 20. Sixty has 12 factors! The idea of factors was not studied by the ancients, but they realised that sharing objects grouped in tens had fewer possibilities than some other grouping schemes. A group of 12 objects could be shared more ways than a group of 10 could, without amounts left over or the need for fractions. Twelve wasn't a popular choice for a number system, but later it was used for measurements that are still evident when we read hour times on a clock. The use of 60 was popular with the Babylonians and is covered in that chapter.

Number systems were also influenced by the materials, resources and technology available at the time. Records needed to be kept, so methods to retain important countable data had to be devised. These methods included marks made in wet clay which was then dried, knots tied a special way on lengths of cord and drawings (hieroglyphics) etched on stone. Many millennia ago, single marks were used to record the information, but these were highly inefficient and took time as well as valuable space. A wedge mark for 10 pushed into clay was faster to make than 10 single

ACTIVITY 3:
What is special about 60?

Go to page 34 and complete the activity to explore the number 60.

Introduction

marks. So real objects or symbols were used to represent groups. These were the first abstractions.

The Maya of Central America drew bars. Each bar had a value of five.

A Sumerian wedge pushed horizontally into clay had a value of 10.

⟨

This ancient Egyptian coil had a value of 100.

Over 4,000 years, cultures in western Asia gradually exchanged ideas on how to manipulate numbers in calculations as trade among people in the region grew. This simple mathematics was difficult to do using symbols like the examples above, so the Arab traders began to use a number system developed by the Hindus in the region of the Indus River in Asia. This system of symbols enabled calculations and so was favoured as it made manipulating numbers in calculations easy. The number system was introduced to Europe about 800 years ago, and since then it has been described as the Hindu–Arabic number system. The Hindu contribution, which occurred long before the Arab modification, is discussed in this book.

As mentioned earlier, fractions were a difficult concept for the ancient civilisations to understand; however, the ancient Egyptians were able to devise a system for expressing fractions, which allowed sharing. This is the first evidence of early fractions, and those early ideas and notations are discussed in this book.

Geometry

ACTIVITY 4:
A famous 3D object

Go to page 35 and
complete the activity.

The most evident geometric aspects of ancient civilisations are the impressive structures that were built thousands of years ago and are still standing today. Usually, these take the form of pyramids with square bases; they were used as temples or monuments for ceremonial purposes or as tombs. A square base was easier than other base shapes to lay out with the technology available, and a building such as a pyramid that tapers to a point (apex) is stable and can withstand extreme forces.

Usually, the first questions about the large ancient structures relate to the technology people used to construct them. Some were built by carving and moving massive stones long distances (Egyptian pyramids and Inca monuments). Others

These Kushite pyramids are found in modern Sudan along the Nile River. They are smaller, steeper and more recent constructions than those found in Egypt. Their name comes from the ancient Nubian Kingdom of Kush, which rose to power after the 24th Egyptian dynasty fell around 3,000 years ago. The Kushites became the leading power in the middle Nile for nearly 1,000 years and for a short time ruled all of Egypt.

were made with small carved stones or manufactured sun-baked bricks that were carefully laid (Babylonian and Indus). Mortar was used sometimes to cover over the irregularities of the large stones or to hold together bricks. Some of the best preserved structures are those built by the Inca in Peru; the stones were carved and fitted together perfectly without mortar.

The Sumerian Ziggurat of Ur, in modern-day Iraq, was constructed over 4,000 years ago. It is a temple whose name implies it has a special aura. After 1,500 years, the temple had crumbled and reconstruction began. This process has continued (on and off) up to the modern era.

Many of the structures were oriented perfectly in certain directions (such as north–south) or, like the pyramids of the Maya, to capture events such as a solstice or an equinox at certain times of the year. The layout of these large buildings prompted the investigation of angles – particularly those that gave square corners or right angles. The oldest preserved written records are facts about right-angled triangles used by the early engineers. Data on the clay tablets or early hieroglyphic markings gave the lengths of each side of right-angled triangles, which were then used to ensure that the corners of the pyramids were square. For example, this rope could be arranged so each length is a side of a right-angled triangle.

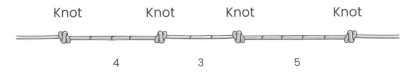

Knot	Knot	Knot	Knot
4	3	5	

Measurement

Information about measurements and measuring in ancient cultures is scant. The level of sophistication varied widely from one civilisation to the next and ranges from virtually no evidence of measurement being used by the Inca to evidence of a more formal system with related units developed by the Hindus.

Ancient civilisations would have measured similar attributes but used different units. The length of rope shown on page 29 was marked in equal intervals, so the idea of repeating a unit of length to describe a measurement was understood. However, the standard used to determine one unit is not clear. A body part such as the foot could have been the benchmark for one base unit, as it was in more recent (but still historical) times. But whose body part was to be used? It is likely that the base unit came from the ruling individual, such as the pharaoh or emperor. That meant regular changes had to be made to the standard base unit depending on who was in charge.

We have evidence that at least one of the early civilisations used measurement tools and units for length, capacity, weight and angles. However, nothing exists to suggest that there was any connection between attributes, such as capacity and weight, like we have in our decimal system today.

Introduction

Probability

Games and pastimes were a part of all ancient cultures. Many of these activities required random number generators, such as dice. The first dice-like objects were bones with possibly only two outcomes. There is evidence that early cultures developed dice like those we use today (with six faces). Archaeological evidence also exists for other dice shapes, often including a 20-face dice (an icosahedron).

Objects such as bones or more regular shapes have been used for thousands of years to play games or help make decisions. Knuckle bones made from sheep anklebones were some of the first random generators of chance outcomes. By the Roman era, cubes were being used by the legionnaires to pass the time during idle moments. Today the mathematics of chance is a highly developed field.

Activity 1: *Sorting and naming triangles*

1. All triangles have two names. They are named based on the *features of their angles* and the *features of their sides*. Look at the pictures of triangles below. Redraw each triangle into the space in the grid where it belongs. Say the two names for each triangle.

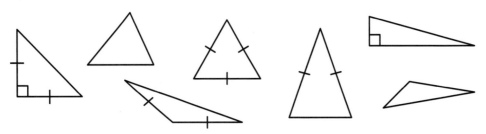

		Names based on angles		
		Acute triangle The biggest angle measures *less than* 90 degrees.	**Right-angled triangle** The biggest angle measures *exactly* 90 degrees.	**Obtuse triangle** The biggest angle measures *more than* 90 degrees.
Names based on sides	**Equilateral triangle** *Three sides* are the same length.			
	Isosceles triangle *Two sides* are the same length.			
	Scalene triangle *No sides* are the same length.			

Activity 2: *All about metric cubes*

In the metric system, cubes are good reference points for the key measurements: length, area, volume, capacity and mass. The three important cubes are shown below. (Note: these cubes are not to scale.) Complete the statements about each of the cubes.

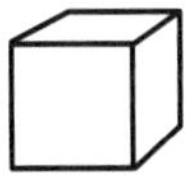

One cubic centimetre or cm³

1. **a.** The length of each edge is one _____.

   b. The area of each face is one _____.

   c. The cube holds one _____ of water.

   d. The mass of water inside is one _____.

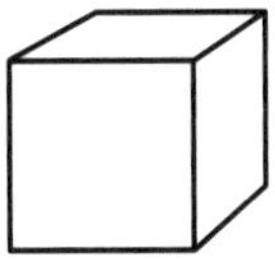

One cubic decimetre or dm³

2. **a.** The length of each edge is one _____ or 100_____.

   b. The area of each face is one _____

   or 100 _____.

   c. The cube holds one _____ of water.

   d. The mass of water inside is one _____.

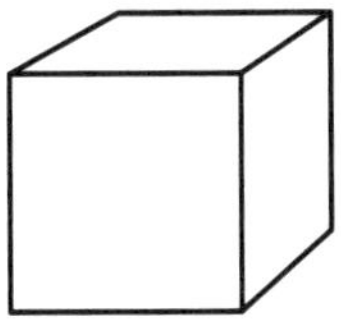

One cubic metre or m³

3. **a.** The length of each edge is one _____ or 100_____.

   b. The area of each face is one _____

   or 10,000 _____.

   c. The cube holds one _____ of water

   or 1,000 _____ of water.

   d. The mass of water inside is one _____

   or 1,000 _____.

Activity 3: *What is special about 60?*

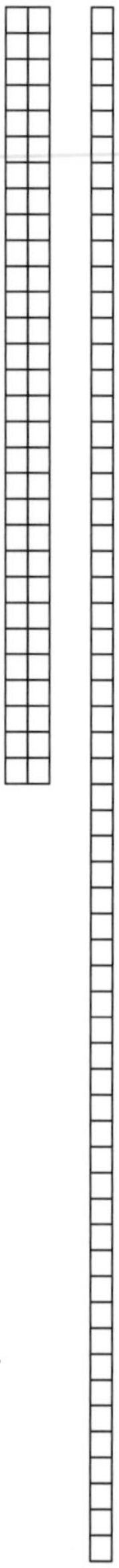

1. The number 60 can be represented as squares arranged in equal columns or rows as shown on the right. What are the dimensions of each of these rectangles?

a. ——— by ——— **b.** ——— by ———

2. Write the dimensions of other rectangles you could draw with 60 squares. You can draw a picture below to help.

a. ——— by ——— **b.** ——— by ———

c. ——— by ——— **d.** ——— by ———

3. Every dimension of a rectangle, like the examples listed above, is called a factor. Write all 12 factors of 60.

_____

> If we divide 60 by a factor, we do not get a remainder or a fraction. That makes 60 an easy number to use in calculations.

4. Sixty is divisible by the one digit factors 1, 2, 3, 4, 5 and 6. Try to find another number less than 100 that has more one digit factors than 60. What did you decide?

_____

Activity 4: *A famous 3D object*

The pyramid was a feature of many early civilisations. The most famous were those constructed in ancient Egypt.

1. A net is a 2D shape of a 3D object after it is unfolded and laid flat. This is a net of a pyramid similar to those built in Giza, Egypt.

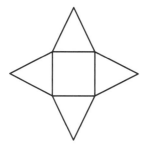

a. What is the shape of the base of this pyramid?

_____

b. The shape of the base is the name, so this is a

_____ based pyramid.

c. What shape are the other four sides?

_____

2. What are the names for each of these pyramids? Think about the base.

a

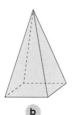
b

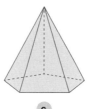
c

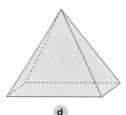
d

a. _____   **b.** _____

c. _____   **d.** _____

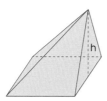

3. Pyramids can have an apex that is shifted to one side. Complete this sentence.

This is a _____ based oblique pyramid.

CHAPTER ONE

Mesopotamia

When? Where? How?
Who? Why?
I wonder, wonder,
wonder, wonder.

What 4,000 year old mathematics
do we use every hour of the day?

The first known written records were
mathematical. What did they say?

How do you create written records
that last for more than 4,000 years?

Why would a civilisation purposefully
choose to use a number system that
was not based on 10?

How long was a Babylonian month?

Who first explored the famous rule
about right-angled triangles?

Mesopotamia:
early evidence of today's mathematics

The oldest recorded evidence of a systematic use of mathematics comes from Mesopotamia where the Sumerian and then the Babylonian civilisations resided along the Tigris and Euphrates river systems. The word *Mesopotamia* means "between two rivers". The first settlements began at, and to the south of, Babylon. For that reason, scholars describe the culture as Babylonian.

M
ACTIVITY 5:
Where are we?

Go to page 56 to study the map and learn some modern names for Mesopotamia.

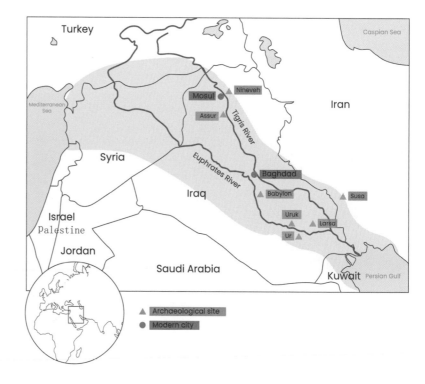

Mesopotamia

This map shows that the river systems lie predominantly in modern-day Iraq and Syria, but they begin in the mountains of Turkey and flow south-east to the Persian Gulf. At one time the empire of Babylon reached west to the Nile River in Egypt and east to the mountains of Iran and possibly beyond.

The rich soil and dependable water supply flowing from the mountains in this region facilitated the establishment of villages. About 5,500 years ago, the population began to grow and large cities were created to cater for the population growth. These were physically walled cities constructed with bricks and governed in a highly organised manner. Schools were established to teach the skills, literature and lore of the culture as well as a system of writing, which was only mastered by a few. This is likely the first example of writing by any culture. The writing was done with a stylus marking in wet clay and then baked. The writing involved a mixture of drawn images and abstract ideas. The numbers written in clay were the first recording of abstract ideas.

FAST FACT

The word *Mesopotamia* was created by the Greeks and means "between two rivers" – these being the Tigris and Euphrates rivers. Interestingly, the last portion (*potamus*) of the word *hippopotamus* also means river. Hippopotamus translates directly as "horse river", suggesting that the animal was thought of as a horse living in the river.

Cuneiform writing is over 5,000 years old and involved over 600 different characters. Each character was a syllable of a word. So a word such as "barley" was broken into "bar" and "ley". A different character was used for each part of the word. Numbers were treated in the same way using a type of place value system.

Numbers

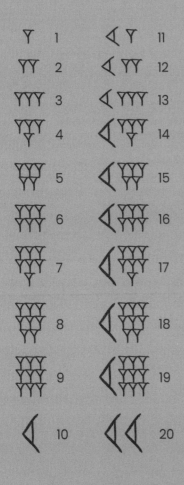

The Babylonian number system used groups of 60 rather than groups of 10 as we use today. Ordinarily, this would mean they needed 60 different symbols to show "how many" in each place. This would have been challenging to remember, so the Babylonians used their stylus to make a *wedge pointing down* to represent one object. A *wedge pointing to the left* meant a group of 10 objects. So one wedge pointing to the left was equal to 10 wedges pointing down.

The wedges were repeated in various combinations to make all numbers from one to 59. This chart shows how they wrote the numbers from one to 20.

A place-value system means that the value of a symbol depends on the place where it is written. The number written in a place is multiplied by the value of the place to give its worth. These individual values must then be added together. So, in our decimal system, the number 235 is made up of 2 in the hundreds place, 3 in the tens place and 5 in the ones place. This can be represented as $2 \times 100 + 3 \times 10 + 1 \times 5$, which you may know as expanded notation.

If the same digits are used in a sexagesimal system (base 60), the value of the number written as 235 is entirely different. It is $2 \times (60 \times 60) + 3 \times 60 + 5 \times 1 = 7{,}385$.

The tables below give the maximum values that can be written for numbers with one, two and three places in decimal and sexagesimal number systems.

Base 10 – decimal number system

Number of places in the number	Greatest value possible
1	9
2	$9 \times 10 + 9 = 99$
3	$9 \times 100 + 9 \times 10 + 9 = 999$

Base 60 – sexagesimal number system

Number of places in the number	Greatest value possible in each place (written in base 10 figures)
1	59
2	$59 \times 60 + 59 = 3{,}599$
3	$59 \times 3600 + 59 \times 60 + 59 = 215{,}999$

At first, the Babylonians did not have a zero, so they just left a space if there was no value in a place. For example, the Babylonians would have written our modern number 3,640 as one group of 60 by 60 plus no groups of 60 plus 40 ones, as shown to the right in Babylonian wedges.

Without a symbol for zero, the sixties place was just left empty, which was obviously very confusing. Much later in the development of the number system, the Babylonians did introduce a symbol to fill the space, but they did not interpret this as the zero we understand today. The idea of *nothing* was difficult for the Babylonians to understand.

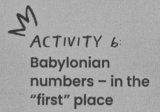

ACTIVITY 6:
Babylonian numbers – in the "first" place

Go to page 57 and complete the activity to decode and code with Babylonian numbers.

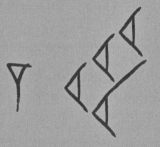

ACTIVITY 7:
Babylonian numbers – in the "first two" places

Go to page 58 and complete the activity to write numbers greater than 60 using Babylonian numerals.

Why use 60 for a base?

There are likely several reasons for the choice of 60 for a base. The first has already been discussed and relates to the fact that the Babylonians, like most early civilisations, could not understand complex fractions. If the base of the number system is 60, it is possible to divide one group of 60 between 2, 3, 4, 5, 6, 10, 12, 15, 20, 30 or 60 without remainders. Multiple groups of 60 makes the task of dividing even easier.

A second possible mathematical explanation for a base of 60 relates to calculations involving the number of days of a year. The Babylonians were good astronomers and kept accurate records of this cycle shown in the sky. They observed that one year could be broken into six sections of 60 days plus a few days. If these few days were ignored during the darkest (shortest) days of the year, important events during nearly all of the year could be predicted using a number system that was the same number as the length of a Babylonian "month". The number 360 (6 × 60) has the advantage of having many factors, which relates to the first point about the choice for 60 as a base.

A third reason why 60 and 360 are prominent in the mathematics of the ancient Babylonians relates to geometry.

Geometry

The fact that a circle is easy to draw meant that the Babylonians likely explored this shape more than any other, except possibly the triangle. The Babylonians viewed the cycle of one year as a circle. In this case, each day is a *step* around that circle. Because it was easy to mark six equal-length arcs around the circle using the length of the radius, it is likely that the circle was drawn around a hexagon. By connecting each vertex (point) to the centre of the circle, they could make six equilateral triangles as shown below. If the angle of each triangle at the centre of the circle is assigned a measure of 60 steps (from the base of their number system), the total around the centre is 360 steps. The fact that 60 in the Babylonian number system could be related to 360 around a circle would have been a very confirming observation. These two numbers were both meant to be and must be used! This may have been one of the greatest "aha" moments in mathematical history.

60 steps

ACTIVITY 8:
The circle and the number 360

Go to page 59 and complete the activity to explore how to construct a hexagon and equilateral triangles in a circle.

ACTIVITY 9:
Name that triangle

Go to page 60 and complete the activity to find triangles you might see in the world where you live.

The relationship between the circle and the six inscribed triangles indicates that the Babylonians might have started their exploration of triangles using what we call equilateral triangles. There isn't any evidence of particular names the Babylonians might have used for triangles, so the words we use here, and throughout this book, come from the Greeks or Romans.

There are a whole range of observations that can be made using the six equilateral triangles inscribed inside the circle. For example, by joining selected vertices, it is possible to construct a right-angled triangle with the hypotenuse sitting on the diameter of the circle. The Babylonians may have observed what the Greeks later proved: that every triangle inscribed in a circle with the hypotenuse on the diameter forms a right-angled triangle.

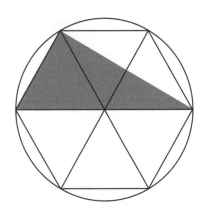

Triangle triples

T he evidence for the knowledge the Babylonians possessed about triangles appears on the famous Plimpton 322 tablet (shown below), which was discovered in the ancient city of Larsa, part of modern-day Iraq. On the tablet is a list of triangle triples. The tablet is estimated to be 3,800 years old. It was donated to Columbia University, New York, in the 1930s and is the subject of considerable research. What information the tablet conveyed wasn't fully understood until 1945 when archaeologist Otto Neugebauer deciphered it. The cuneiform writing is simply a list of sets of three numbers that form right-angled triangles.

The triples of numbers on this tablet would have been found by observation. There is no evidence that the Babylonians had a rule for the relationship between the lengths of the sides of a right-angled triangle. The Greeks discovered the rule to square the numbers as they worked in practical situations such as tiling floors.

The study of triangles was very important to the Babylonians. Some of this can be associated with the triangles drawn inside hexagons as described in the previous section. There is also verifiable evidence that they knew about the relationship between the three lengths of the sides of special triangles (called right-angled triangles today). Today the three numbers that form right-angled triangles are called Pythagorean triples. The name comes from the Greek mathematician Pythagoras who likely learned about the relationship from his studies in Egypt some 2,500 years ago. As you may know, Pythagoras (or one of his followers) also proved the relationship that exists between the sides of a right-angled triangle. The formula, involving algebra, was formalised even later. Interestingly, the mathematics used to establish the formula relating the three sides of a right-angled triangle was developed by Persians who resided in the Mesopotamia region after the Babylonians.

Each right-angled triangle listed on the tablet has different *whole number* lengths – triangles like these with three unequal sides are called scalene triangles. The list does not include any isosceles right-angled triangles (with two sides the same length). This is partially due to the fact that isosceles right-angled triangles can only be constructed with *messy* numbers. These numbers are those pesky *irrational* numbers that cannot be written as fractions!

The particular right-angled triangle examples given by the Babylonians are today called Pythagorean primitives. These are the first three numbers in a sequence that make right-angled triangles. For example, triangles with lengths that have a relationship of 3:4:5 form a right-angled triangle. That means multiples of these numbers, such as 6:8:10, 9:12:15 and 12:16:20, also make right-angled triangles.

ACTIVITY 10:
Right-angled
triangle triples

Go to page 61 and complete the activity to explore Pythagorean primitives that the Babylonians used to work with right-angled triangles.

As the chart below shows, an endless number of right-angled triangles can be made with the primitive 3:4:5. Once the Babylonians knew a primitive, they could "multiply" to increase each length by the same amount to produce a new right-angled triangle.

Right-angled triangle multiples

3 : 4 : 5
6 : 8 : 10
9 : 12 : 15
12 : 16 : 20
.
.
3n : 4n : 5n

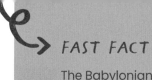

Measurement

The Babylonians could work with triangles with various types of angles, but for right-angled triangles they were limited to triangles with certain side lengths. They used whole numbers to describe the lengths of sides of triangles. However, there isn't clear evidence about the length units that might have been used. Again, it is likely that body parts provided the main units. Some cultures based their units of length on the *cubit*, the distance from the elbow to the end of the longest finger. This convenient length was broken into shorter units that were also based on body parts – the foot, hand or finger. Longer units could have included a step or pace as was used in ancient Rome. If the cubit was used, the reference point for it was very often a king or priest.

The Babylonians did *calculate* the area of regions. The example in this photograph shows the summary of work by Babylonian surveyors to calculate the area of a trapezoidal shaped field. The translation indicates they multiplied the average of the two lengths by the average of the two widths. It is a clever method for this early civilisation and is very close to the formula used today.

The area of this trapezoid was likely calculated using an averaging method and is written inside the shape. The digits in the base 60 system are 5, 3 and 20. The 5 is a whole number and the other two are fractions. Today, a semicolon (;) is used to separate the whole number from the fraction part of the answer so we write this as 5 ; 3 20 in the sexagesimal system. Therefore, the total area is 5 plus 3 × (1/60) plus 20 × (1/3,600).

The object in the image on the next page is nearly 4,000 years old and shows some facts the Babylonians connected to the circle. The argument suggests that the Babylonians only used 3 as their value for pi, which was satisfactory for constructions and probably easier to use in calculations involving their number system. However, further translations of tablets found at Susa have indicated that the Babylonians did have a more accurate estimate of pi as approximately 3.125. The thinking is very similar to what is used today, although the value for π differs from what is now known to be the accurate value.

This tablet was excavated in 1936 by French archaeologists in the ancient city of Susa, located in modern-day Iran.

The Babylonian numeral 3 above the circle on the tablet above seems to indicate the circumference of the circle, which likely has a diameter of one unit, and hence be their estimate for pi. This conclusion has been made based on the following argument.

Two rules – Circumference $(C) = 2\pi r$ and Area $(A) = \pi r^2$ – have been used to interpret the following.

Square both sides of $C = 2\pi r$ to obtain $C^2 = 4\pi \times \pi \times r^2$.

Now substitute A into the equation $C^2 = 4\pi \times A$.

The sexagesimal fraction 45 is written in the circle and 3 is written above it, so assume that $A = \dfrac{45}{60}$ and $C = 3$ and replace these values in the equation.

$3^2 = 4\pi \times \left(\dfrac{45}{60} \right)$ or

$9 = 4\pi \times \left(\dfrac{3}{4} \right)$ or

$9 = \pi \times 3$.

This means that the Babylonians estimated pi to have a value of 3.

Begin with the equation $C = 2\pi r$ for the circumference, which the Babylonians likely knew, and write

$$3 = 2\pi r$$

or $\dfrac{3}{2\pi} = r$

Using the equation for the area of a circle, $A = \pi r^2$, and substituting the value above for r, the equation becomes

$$A = \pi \left(\frac{3}{2\pi} \right)^2$$

This result can be simplified to $\dfrac{9}{4\pi}$.

The numeral in the picture inscribed inside the circle is 45 in the Babylonian base-60 system. This is equivalent to 0.75, which is exactly the same as 9 divided by 4π, assuming the Babylonians used a value of 3 for pi.

The most interesting aspect of the Babylonian measurement system was the contribution that the culture made to measuring angles. The Babylonian words for the unit used to describe angles is "step" or "step down". In Latin the step down is translated as *de-gree*. So the modern-day use of the word *degree* can be traced directly back to the Babylonians.

The use of 360 little steps or degrees around a point is the basis for much of the geometry that we use today (think of a protractor). The fact that this unit has been used for nearly 4,000 years and is now embedded in both Euclidean and non-Euclidean geometry is impressive.

The Babylonian work with measurement of angles would have influenced the development of time. Before the digital age, almost all clocks displayed 12 hours and 60 minutes marked around the circle of time. Many clocks had two or three *hands* (hour, minute and second hands). These hands

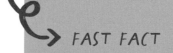

Expressions with exponents, such as $(x + y)^3$ or z^4, are described using the word "degrees". For the examples here, we say that $(x + y)$ is raised to the third degree or that z is raised to the fourth degree. Something that is limitless might be called the *nth degree*. This language is appropriate because raising an expression to a power is like taking steps up or down a ladder.

3rd

2nd

1st

would consistently cycle past the marks, so 12 hours were passed during a *day*, or more specifically, daylight. The other hands showed that one hour was divided into 60 minutes, which in turn were each broken into 60 seconds. This made it easy to calculate fractions of time. Today's use of 60 minutes and 60 seconds on a clock came from the Babylonian practice of breaking degrees into these same fraction parts.

The Babylonians' work with the circle provided a basis for both hours and minutes. If the six central angles formed by corners of equilateral triangles were each cut in half, the circle was divided into 12 parts (12 hours) as shown below.

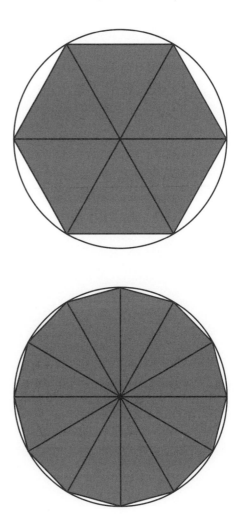

Babylonian astronomers had already identified the 12 significant celestial groupings of stars, so this was a natural choice for the number of parts around a circle. The Babylonians used a tall obelisk like the picture below. It cast a shadow to show the progression of time throughout the day. But this is not a sundial. Sundials were first used in Egypt, not Mesopotamia.

The Greek historian Herodotus wrote that the Greeks adopted the use of sundials and dividing a day onto 12 hours from the Babylonians. The division of a day into 12 hours was also adopted by the Romans, continuing the Babylonian legacy. To divide the day into 12 parts, it would have been necessary to have a time-measuring device. However, this time-measuring tool has been ellusive to identify. The earliest methods, at least in Egypt, were obelisks. However, to date, only a very few obelisks have been found in the Tigris–Euphrates region of the Middle East. The most famous is this tall stele, an obelisk-like structure. It could have been used to mark the progression of time throughout the day.

This obelisk is located at Nimrud along the Tigris River in modern-day Iraq. It is a memorial to commemorate the deeds of King Shalmaneser III. For that reason, it could (or should) be called a *stele*. The word *obelisk* comes from the Greek *obelus*, which was a punctuation mark used in ancient manuscripts to denote that a word was questionable. The shape of the mark is like the division sign in mathematics (÷). The reason for this similarity is unclear.

ACTIVITY 11:
Babylonian mathematics word search

Go to page 62 and complete the word search activity.

Summary

Mesopotamia was located between and around the Tigris and Euphrates rivers in modern-day Iraq and Syria. It was usually described as Babylon. The first mathematics has been dated to 5,400 years ago.

Numbers

- Used a base-60 system (sexagesimal) to group numbers
- Did not initially use a zero
- In some of the first written records, denoted numbers by wedge marks

Geometry

- Explored six equilateral triangles inscribed in a circle
- Assigned 60 as a measure at each corner of an equilateral triangle
- Knew the relationship between the length of sides of a right-angled triangle

Measurement

- Assigned 360 "steps" as the total measure of all angles around a point
- Used an averaging method to calculate the area of shapes
- Separated daylight into 12 units

Algebra

- Probably used generalisation techniques, as suggested by their methods of dealing with right-angled triangles

The Babylonian legacy

The Babylonians used degrees based on 360 degrees around a point and used 60 as a measure for minutes and seconds. They knew about sets of three whole number lengths for right-angled triangles.

Activity 5: *Where are we?*

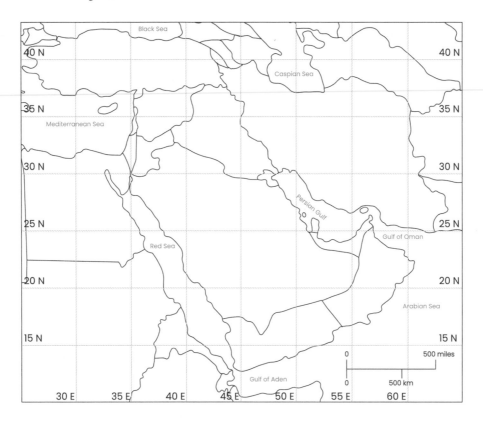

1. Use blue to shade these bodies of water on the map.

    a. Mediterranean Sea   **b.** Red Sea   **c.** Caspian Sea

    d. Black Sea   **e.** Arabian Sea   **f.** Persian Gulf

2. Use these clues to write the country names on the map.

    a. Egypt – just to the west of the Red Sea

    b. Saudi Arabia – just to the east of the Red Sea

    c. Iran – just to the north of the Persian Gulf

    d. Iraq – just to the west of Iran, with very small access to the Persian Gulf

    e. Turkey – just north of Iraq and west of Iran.

3. Babylon was located in modern-day Iraq between these two rivers. The rivers both started in Turkey and emptied into the Persian Gulf. Sketch and label where these two rivers could have flowed.

    a. Tigris River   **b.** Euphrates River

Activity 6: *Babylonian numbers – in the "first" place*

1. Write the Hindu-Arabic numeral and the number name for each of these Babylonian wedge marks.

a. 《《 ΥΥΥ _____

b. 《《《 ΥΥ _____

c. (wedge marks) _____

d. (wedge marks) _____

The numbers you write are called Hindu-Arabic numerals. They are much more recent than the Babylonian numbers.

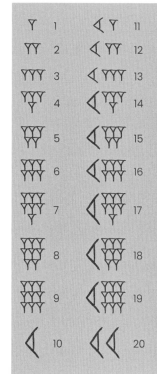

Υ	1	《 Υ	11
ΥΥ	2	《 ΥΥ	12
ΥΥΥ	3	《 ΥΥΥ	13
ΥΥΥΥ	4	《 ΥΥΥΥ	14
ΥΥΥΥΥ	5	《 ΥΥΥΥΥ	15
ΥΥΥΥΥΥ	6	《 ΥΥΥΥΥΥ	16
ΥΥΥΥΥΥΥ	7	《 ΥΥΥΥΥΥΥ	17
ΥΥΥΥΥΥΥΥ	8	《 ΥΥΥΥΥΥΥΥ	18
ΥΥΥΥΥΥΥΥΥ	9	《 ΥΥΥΥΥΥΥΥΥ	19
《	10	《《	20

2. Draw Babylonian wedge marks for these Hindu-Arabic numerals.

a. 25 _____   b. 52 _____

c. 29 _____   d. 30 _____

e. 34 _____

3. What is the greatest number of 《 that would be written in the "ones" position of the Babylonian number system? _____

Because the Babylonians work with groups of 60, they need show up to 59 in the ones position.

4. What is the greatest number of Υ that would be written in the "ones" position of the Babylonian number system? _____

Activity 7: *Babylonian numbers – in the "first two" places*

1. Write the Hindu–Arabic numeral and the English number name for each of these Babylonian wedge marks.

 The Babylonians did not have a symbol for nothing (zero). They just kept the position empty.

Sixties	Ones	Numeral	Number name
a. 𒁹	𒁹𒁹𒁹	$1 \times 60 + 3 =$ _____	
b. 𒁹𒁹𒁹	𒌋𒌋𒁹	$3 \times 60 + 21 =$ _____	
c. 𒌋	𒌋𒌋𒁹𒁹𒁹	_____ $\times 60 +$ _____ $=$ _____	
d. 𒌋𒌋𒁹		_____ $\times 60 +$ _____ $=$ _____	

2. Draw Babylonian wedge marks for these Hindu–Arabic numerals.

a. 75	**b.** 80
c. 100	**d.** 200

3. Use Babylonian wedge marks to write the greatest number possible for a number written using both the sixties and ones positions.

_____

4. Write the number in Question 3 using Hindu–Arabic numerals.

_____

Activity 8: *The circle and the number 360*

Follow these steps to explore the connection between a circle, a hexagon and then equilateral triangles.

Step 1 Use the compass to draw a circle.	**Step 2** Keep the compass arms spread to match the radius. Mark around the circle. This will give six equally spaced intervals. Number the marks as shown.

Step 3 Now use a straight edge to draw lines to connect the numbers on the circle.	**Step 4** Connect the vertices (the points you have numbered) to the centre of the circle.
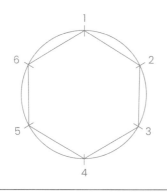	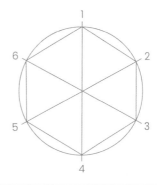

1. Complete these statements.

    a. The shape made by connecting the six numbers on the circle is a

    regular _____.

    b. The six identical shapes are called _____ triangles.

Activity 9: *Name that triangle*

1. Look around. Where do you see triangles? You may need to go outside or think about places you have recently been.

_____

_____

_____

_____

_____

_____

> Triangles provide support to strengthen all structures. You can see the shapes here but many buildings hide the triangles that are used to give stability.

2. There are two types of right-angled triangles shown below.

a. Write any other names you know for each of these right-angled triangles.		**b.** Draw pictures of two triangles that *do not have* right angles. Write any special names you use for these triangles.	
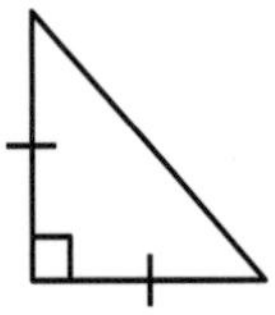			

Activity 10: *Right-angled triangle triples*

 The Babylonians listed sets of three numbers that made a right-angled triangle. The activities on this page show that every triple could be used to find many other right-angled triangles.[1]

The rule for determining whether the numbers form a right-angled triangle is as follows: multiply each number by itself and check that the sum of the two smaller numbers equals the third number. See example below

	$(3 \times 3) + (4 \times 4) = (5 \times 5)$		
3 : 4 : 5	9	+ 16	= 25

1. Write the missing numbers for the right-angled triangle rule that shows how these number triples are related.

a. 3 : 4 : 5 → 9 + 16 = _____	**b.** 6 : 8 : 10 → 36 + _____ = _____
c. 9 : 12 : 15 → ___ + ___ = ___	**d.** 12 : 16 : 20 → ___ + ___ = ___

2. Multiply each of these numbers 3 : 4 : 5 by 10. Write the results.

_____ : _____ : _____

Show your calculations here to check that this is a right-angled triangle.

3. Check that the first three numbers in each list form a right-angled triangle. Then double, triple and quadruple the first set of numbers to create more right-angled triangles. The first list has been started for you.

	5 : 12 : 13 $5^2 + 12^2 = 13^2$ 25 + 144 = 169 Yes, this forms a right-angled triangle	7 : 24 : 25	9 : 40 : 41
Double			
Triple			
Quadruple			

[1]Research for this is based on work done at the University of New South Wales in 2017.

Activity 11: *Babylonian mathematics word search*

1. Find these words that are hidden in the grid below. All words are written horizontally left to right, vertically top-down or diagonally top-down (sloping to the right or to the left).

BABYLON   ANGLE   PERSIAN   TIGRIS   WEDGE

SIXTY   TRIANGLE   HINDU   ZERO   FACTOR

ASTRONOMERS   PYTHAGOREAN   ARABIC   EUPHRATES   RIGHT

CIRCLE   MESOPOTAMIA   NUMERALS   STYLUS   TABLET

DEGREE

A	N	M	D	R	C	J	T	F	K	R	F	S	B	A	D	L	B	U	S
C	D	U	A	N	O	I	P	H	G	P	O	A	W	E	P	A	T	Q	F
Y	G	Y	M	F	V	Y	R	G	Y	A	S	D	L	B	O	T	W	M	T
O	U	B	I	E	T	O	Z	T	A	T	I	J	L	W	E	D	G	E	H
D	S	L	F	H	R	Y	H	Q	U	H	A	A	R	I	Y	T	A	S	O
K	B	C	Y	U	M	A	U	V	B	Y	O	B	S	D	V	A	F	O	I
L	E	D	N	G	G	Q	L	A	S	D	L	P	A	G	K	R	E	P	S
C	X	I	D	O	L	G	I	S	U	Y	E	N	S	I	X	T	Y	O	B
E	G	W	R	B	E	A	S	A	C	I	G	R	T	G	N	I	Y	T	R
O	T	E	R	A	D	R	U	D	A	L	A	Z	R	N	O	L	K	A	T
C	A	E	Y	B	A	F	G	T	E	L	O	W	O	T	U	B	J	M	C
N	W	P	I	Y	N	N	S	E	P	G	A	F	N	S	A	I	M	I	C
C	I	R	C	L	E	D	U	X	O	R	R	B	O	H	E	B	I	A	P
H	V	S	Y	O	S	P	T	Y	I	N	O	E	M	A	R	A	L	P	K
I	E	P	T	N	H	K	L	T	L	A	T	S	E	H	V	Z	N	E	J
L	D	A	C	R	Y	D	T	A	E	V	P	A	R	T	R	M	E	K	T
Z	J	M	A	V	W	O	D	C	O	P	E	R	S	I	A	N	Y	R	W
K	E	T	O	R	N	T	J	B	K	M	P	A	W	D	U	L	R	G	O
V	E	I	F	L	I	R	V	K	Q	F	C	B	R	M	Y	R	I	M	K
S	R	X	E	G	H	I	N	D	U	D	A	I	G	P	W	J	G	O	G
C	O	F	R	E	W	A	X	C	R	O	S	C	R	R	E	N	H	R	A
H	T	I	F	M	T	N	E	Y	O	M	T	H	T	F	M	U	T	F	G
A	S	T	P	A	R	G	I	A	F	R	A	B	G	O	S	E	B	Y	O
D	L	F	Q	N	Y	L	S	N	D	E	P	R	A	D	R	O	A	C	R
F	G	E	D	H	L	E	A	S	T	Y	L	U	S	E	W	F	W	I	E

Working space

CHAPTER TWO

The Egyptians

When? Where? How?
Who? Why?
I wonder, wonder,
wonder, wonder.

Where is the only structure surviving of the original Seven Wonders of the Ancient World?

What is the longest river in the world?

How many faces does the Great Pyramid of Giza have?

Why was the Egyptian way of writing fractions unique?

How did the Egyptians measure right-angles?

The Egyptians:
giant geometric achievements

Egypt is one of the most visited and widely talked about places in the world, and it is an archaeological treasure trove. There is a lot of information about the cultural and architectural history of the Egyptian civilisation. One of the reasons for this is that the history has been very well preserved in the pyramid tombs of pharaohs and by Egypt's warm, dry climate.

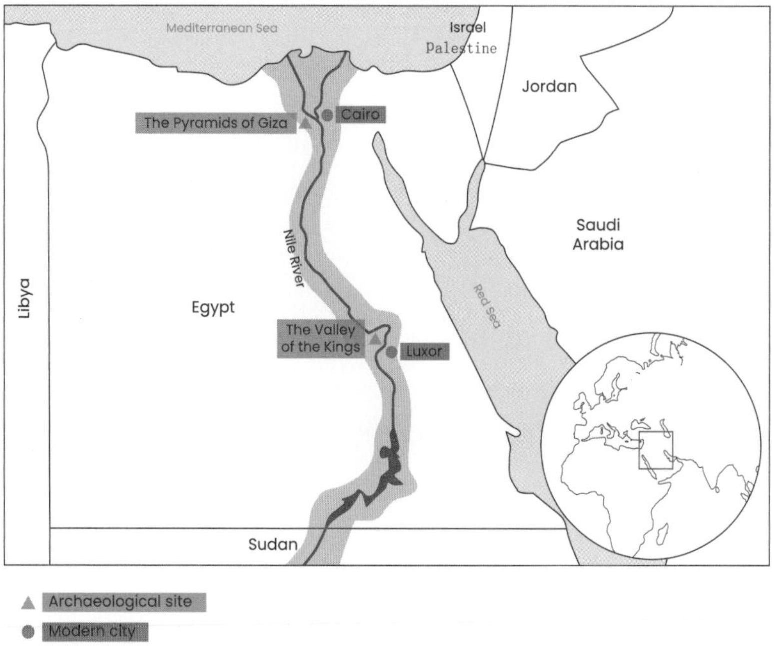

ACTIVITY 12:
Where in the world is Egypt?

Go to page 92 and complete the activity to discover and explore the world of ancient Egypt.

Mediterranean Sea
Israel
Palestine
Jordan
The Pyramids of Giza
Cairo
Libya
Nile River
Egypt
Saudi Arabia
Red Sea
The Valley of the Kings
Luxor
Sudan

▲ Archaeological site
● Modern city

Egypt is in the north-eastern corner of the African continent. The warm climate and the abundant water supply from the Nile River meant that people could prosper, and they developed agricultural techniques that sustained a growing population. The civilisation relied heavily on agriculture, so they needed to maintain systematic records of all sorts of information. This included marking land space for growing crops, measuring amounts of grain for commercial sale and tracking time to understand and predict seasonal changes. For each of these tasks, the Egyptians relied on their mathematical understanding.

The Nile is the longest river system in the world. Its length of 6,650 km forms a basin for 11 countries: Ethiopia, Eritrea, Sudan, Uganda, Tanzania, Kenya, Rwanda, Burundi, Egypt, Democratic Republic of the Congo, and South Sudan.

There are four river systems longer than 6,000 km.

River	Continent	Length
Nile	Africa	6,650 km
Amazon	South America	6,400 km
Yangtze	Asia	6,300 km
Mississippi	North America	6,275 km

The Egyptian number system

The Egyptian number system provides an interesting contrast to the Babylonian place-value system for recording numbers. The earliest documented evidence of the Egyptian number system came approximately 5,000 years ago, which is about the time the Babylonians first recorded numerals by pressing wedges into wet clay. Rather than using durable clay, the Egyptians made markings on papyrus reeds that grew along the banks of the Nile River. Papyrus was well suited to the arid Egyptian climate – it could be rolled up, carried and stored with ease. As the early Egyptian civilisation developed, the people needed to maintain systematic records of information to ensure the population thrived.

The main number system used by the ancient Egyptians was based on pictorial symbols known as hieroglyphics. There were seven symbols for numbers – each one represented a different power of 10 from one through to one million. These seven symbols were then used in a tally system to represent greater numbers. Each symbol had a particular meaning based on an aspect of Egyptian society and culture. Below is a list of Egyptian number symbols and their values.

Historians have long debated the meaning behind these seven symbols. The first three symbols (1, 10 and 100) are thought to represent ropes of varying lengths and shapes according to their degree of magnitude. The symbol for 1,000 represents a lotus flower that grew in abundance in the Nile River. Ten thousand is a bent finger. The symbol for 100,000 is most likely a stylised tadpole, which lived in abundance in the Nile River. Finally, the symbol for one million is an "astonished" figure with upraised arms, possibly signifying vastness or eternity.

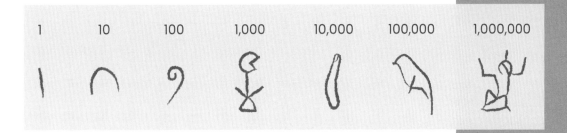

The way that numbers were written in the Egyptian system was purely additive. The number of times a symbol was used determined its collective value. The order or position of the symbols was irrelevant since the Egyptian system was not place-value dependent, nor was there a symbol for zero to act as a placeholder. However, the common practice was to write numbers from left to right in ascending order. You read the numbers as you would read tallies. For example, the Egyptian numeral below is 1,240,358. It is possible the Egyptian priests invented this system about 4,000 to 5,000 years ago.

ACTIVITY 13:
The Egyptian number system

Go to page 94 and complete the activities to "crunch" some Egyptian hieroglyphics and decode the numbers!

ACTIVITY 14:
Decoding Egyptian hieroglyphics

Go to page 95 and complete the activities to decode Egyptian hieroglyphics.

One of the main uses of hieroglyphs was to help Egyptians record information about their population, the yearly harvest or the spoils of war. An example is the ceremonial mace head pictured on the next page, which dates to around 3000 BCE. The inscriptions show the spoils of King Narmer's conquests. The drawings below, of carvings from the other side of the mace head, show oxen, goats and prisoners (with arms tied behind the back), respectively, with hieroglyphs to indicate the number of each captured. Although the numbers are probably exaggerated, it appears he took 400,000 oxen, 1,422,000 goats and 120,000 prisoners.

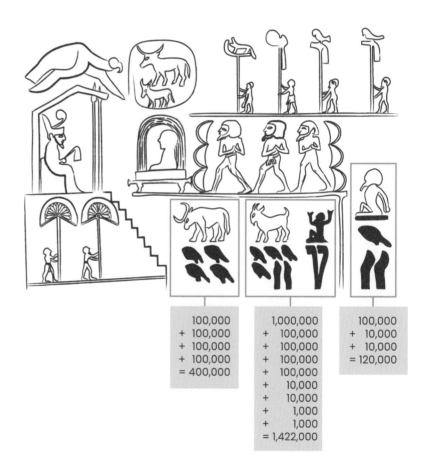

100,000	1,000,000	100,000
+ 100,000	+ 100,000	+ 10,000
+ 100,000	+ 100,000	+ 10,000
+ 100,000	+ 100,000	= 120,000
= 400,000	+ 100,000	
	+ 10,000	
	+ 10,000	
	+ 1,000	
	+ 1,000	
	= 1,422,000	

The Egyptians

A macehead was originally used in combat, but over time became a symbol of power and indicated the success of rulers. This is a macehead credited to the Egyptian pharaoh King Narmer who ruled Egypt about 5,000 years ago. It describes the captured loot from the conquest of a rival nation. Some references cite this as the macehead of Pharaoh Menes. There is now consensus that these two rulers were the same individual. The seal (or serekh) at the top is translated directly as Narmer.

Operations with Egyptian numerals

To organise their commercial activities, the ancient Egyptians needed to work with number operations, such as addition, subtraction and multiplication. While there is not much recorded evidence of Egyptians using addition or subtraction, we assume they would have needed to understand and use these operations for day-to-day record keeping and even to understand their tally-like numbers!

Adding Egyptian numbers would have been relatively straightforward and similar to how we add numbers today. Adding two sets of numbers would simply have involved grouping all like symbols. Where there were 10 or more symbols, a regrouping step would be used. For example, 18 plus 24 would equal 3 tens symbols and 12 ones symbols. The 12 ones would then be regrouped to make 1 tens symbol and 2 ones symbols. The final tally would be 4 tens and 2 ones symbols. The Egyptians may have had a similar system for subtracting numbers, but there is no direct evidence of the method used.

The more interesting mathematical development was the ancient Egyptian method of multiplication. Their method helped them overcome shortcomings in their simple tally-like number system. The ancient Egyptians understood multiplication could be completed relatively easily by doubling. For example, to multiply by 2, they simply doubled. To multiply by 4, they doubled and then doubled again, so 18 multiplied by 4 was "double 18 and then double 36". Today, this thinking is taught in many schools. Below are examples of the doubling method, first using modern-day Hindu–Arabic numerals, then using hieroglyphics.

Doubling using hieroglyphics

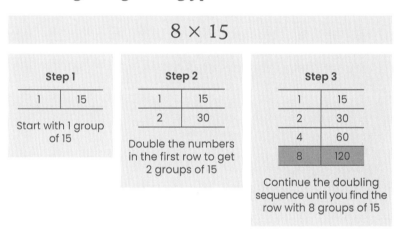

8×15

Step 1	
1	15

Start with 1 group of 15

Step 2	
1	15
2	30

Double the numbers in the first row to get 2 groups of 15

Step 3	
1	15
2	30
4	60
8	120

Continue the doubling sequence until you find the row with 8 groups of 15

Sometimes, the product of two numbers (that is, the result when they are multiplied) could not be found in a single doubling row. In those cases, it was necessary to add numbers in two or more rows to find the final product. For this situation, the Egyptians doubled the number,

ACTIVITY 15:
Multiplication by doubling

Go to page 96 and complete the activities to discover how to use doubling to multiply any pair of whole numbers.

stopping just before exceeding the multiplier. Combinations of the doubles were then selected that totalled the multiplier, as shown in the following example.

10 × 15

Step 1

| 1 | 15 |

Start with 1 group of 15

Step 2

| 1 | 15 |
| 2 | 30 |

Continue until just before exceeding 10

Step 3

1	15
2	30
4	60
8	120

150

Choose and add numbers that equal 10

These examples show the same steps using hieroglyphics. The Egyptian system was easy to use because it was additive.

ACTIVITY 16:
Using Egyptian numbers to multiply

Go to page 97 and complete the activity to multiply using Egyptian hieroglyphics to multiply.

8 × 15

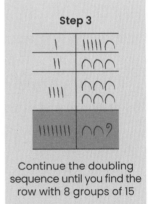

Step 1

Start with 1 group of 15

Step 2

Double the numbers in the first row to get 2 groups of 15

Step 3

Continue the doubling sequence until you find the row with 8 groups of 15

10 × 15

Step 1

|	|||||∩

Start with 1 group of 15

Step 2

|	|||||∩
||	|||||∩ |||||∩

Double both numbers

Step 3

|	|||||∩
||	∩∩∩

Regroup the ones

Step 4

|	|||||∩
||	∩∩∩
||||	∩∩∩ ∩∩∩
||||||||	∩∩∩ ∩∩∩ ∩∩∩ ∩∩∩

Keep doubling the number sequence until just before exceeding 10

Step 5

|	|||||∩
||	∩∩∩
||||	∩∩∩ ∩∩∩
||||||||	∩∩⌒

Regroup the tens

Step 6

|	|||||∩
||	∩∩∩
||||	∩∩∩ ∩∩∩
||||||||	∩∩⌒

Locate the combination of factors/rows that make 10 lots of 15. Highlight them. The sum of the corresponding numbers in the right-hand column is the product of 10 and 15.

It is likely the ancient Egyptians also used the doubling method for division. The process would have been manageable, but at times awkward. It is easy to imagine that problems arose whenever the Egyptians had to deal with numbers that could not be divided evenly. To overcome these problems, the Egyptians had to use fractions. Their understanding and use of fractions is the earliest ever discovered!

An eye for fractions

As far as we know, the ancient Egyptians were one of the first cultures to understand and apply fractions. They used symbols related to the Eye of Horus (a stylised eye symbol) to show commonly used fractions like $\frac{1}{2}$, $\frac{1}{4}$, $\frac{1}{8}$, $\frac{1}{16}$, $\frac{1}{32}$ and $\frac{1}{64}$. As you can see, the Egyptian notation is limited to unit fractions, that is, fractions with a numerator of 1.

For the ancient Egyptians, the Eye of Horus was an enduring symbol of sacrifice and redemption, but it was also highly practical and mathematical. Each piece of the eye was used to represent a fraction of a *hekat*. A *hekat* is one of the oldest Egyptian measures for the volume of grain, beer or bread. If you convert it to modern units, 1 *hekat* is the same value as 4.8 litres.

While it is not apparent why the fractions were allocated their respective symbols, there is general agreement that:

- $\frac{1}{2}$ *hekat* corresponds to the right part of the eye

- $\frac{1}{4}$ *hekat* corresponds to the pupil

The Egyptians

- $\frac{1}{8}$ *hekat* corresponds to the eyebrow

- $\frac{1}{16}$ *hekat* corresponds to the left part of the eye

- $\frac{1}{32}$ *hekat* corresponds to the curling tail which represents taste

- $\frac{1}{64}$ *hekat* corresponds to the tear.

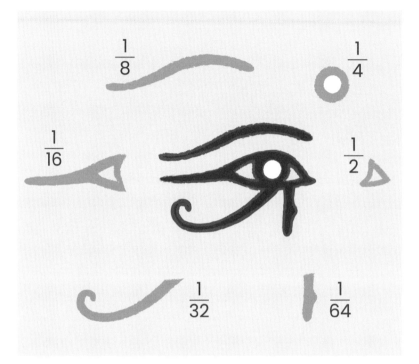

Several observations can be made about these fractions. First, the denominators increase by doubling. This makes it easy to find equivalent values, such as one-half being the same as two-fourths, one-fourth being the same as two of the eighths and so on. This feature does not help overcome the problem of expressing fractions with denominators such as thirds, fifths and so on. The Egyptians may have limited themselves to using just a few fractions. A second observation is the fact that all the fractions listed above add to $\frac{63}{64}$, which is just short of a full *hekat*. Some historians believe that since Thoth replaced Horus's eye, the missing fraction was withheld by his magic. It could also symbolise that nothing is perfect!

ACTIVITY 17:
An eye for fractions

Go to page 98 and complete the activity to explore the fractions used by the Egyptians.

The ancient Egyptians seem to have understood that fractions could come in more advanced forms than the Eye of Horus notation suggested. The Rhind Papyrus, attributed to the Egyptian scribe Ahmes in 1650 BCE, showed that the Egyptians had knowledge of fractions with a numerator of two. They also had special symbols for $\frac{2}{3}$ and possibly $\frac{3}{4}$, but all other fraction notation was limited to unit fractions. These fractions were represented by a flattened oval, representing a numerator of 1, above the denominator. There was no specific notation for writing fractions with a numerator greater than 1, so the Egyptians had to come up with their own method.

The solution involved writing any fraction as the sum of a set of unit fractions with all different denominators. For example, they wrote a fraction such as five-sixths as the sum of one-half plus one-third.

$$\frac{5}{6} = \frac{1}{2} + \frac{1}{3}$$

Sometimes it was necessary to use more than two unit fractions to give the value.

$$\frac{5}{7} = \frac{1}{2} + \frac{1}{5} + \frac{1}{70}$$

The ancient Egyptians tried to be as efficient as possible, so they always started with the greatest unit fraction possible. In the examples above, the first fraction is one-half, which is the greatest possible unit fraction. To find the greatest unit fraction possible, the ancient Egyptians had to estimate and use trial and error until they landed on the correct fraction.

ACTIVITY 18:
Greedy algorithms

Go to page 99 to explore the Egyptian algorithm which today could be classified as "greedy" (a term for a computer algorithm that takes the fewest number of steps possible). In this case, the Egyptians expressed every fraction as a sum using the fewest unit fractions possible.

The Egyptians

As you can see, the ancient Egyptian notation is a long way from the proper and improper fractions we can use today. However, it is still astonishing that they were able to comprehend such an abstract idea!

This is an image of a section of the famous papyrus that was copied by the Egyptian scribe Ahmes about 3,500 years ago. The papyrus is a record of methods that were used to solve mathematical problems. It was purchased by the collector Alexander Rhind in 1858. It is often described as the Rhind papyrus, but in reality, was the work of Ahmes.

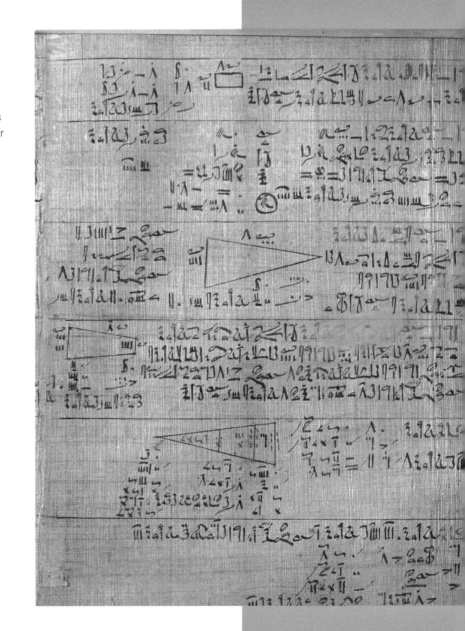

Measurement

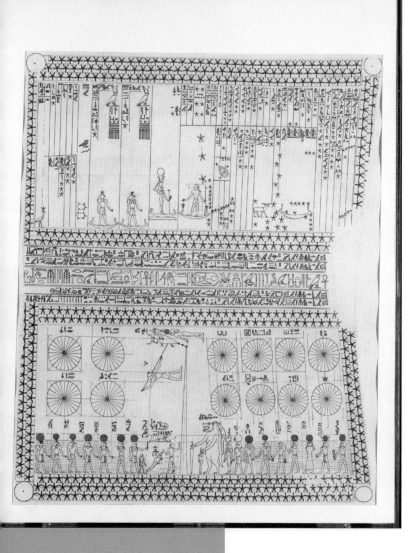

The ancient Egyptians were one of the earliest civilisations to construct and define a timekeeping system to make sense of the world around them. They came up with a series of calendars and ways of measuring time based on sun and moon cycles.

This image was found (aptly) on the ceiling in the tomb of Senemut, an astronomer to Pharaoh Hatshepsut. It represents a year of 12 months of 30 days each plus an extra five or six days. The illustration is likely 4,700 years old and shows constellations from that time. The 13 images with red headgear represent moons – about one out of every three years has 13 full moons.

The Egyptians

One of the main reasons why the Egyptians needed a calendar was to determine how the change of seasons affected crop production. They found that the annual flooding of the Nile came at around the same time as the reappearance of a bright and distinctive star in the eastern sky that was known to the Egyptians as Sirius. By counting days, the Egyptians calculated that the reappearance of Sirius happened only once every 365 days. They used this event to mark the beginning of a new solar year. They composed a calendar that was broken into four main seasonal cycles of 90 days each, plus five additional days.

While the Egyptian estimate of a year was remarkably close to the measure we use today, there was a slight miscalculation. The Egyptian calendar did not run for a complete solar cycle but fell behind by about one day every four years. Eventually, this threw the Egyptian calendar out of alignment with the true astronomical year. One Egyptian king tried to correct this through the introduction of an additional calendar day every four years; however, his proposed decree was resisted by the Egyptian priests and people. The proposition was abandoned until the time of the Romans, who noticed a similar discrepancy with their calendar. It is not clear what the Egyptians did to correct the misalignment in their calendar. It is possible that they simply restarted the calendar with the annual rising of Sirius. It could also be that they just learnt how to live with a calendar that deviated from the cycles of nature.

Before clocks and watches were invented, people relied on patterns of day and night to tell time. One of the earliest devices for recording time was called a sundial. As its name suggests, the sundial was used to record the passage of hours based on shadows cast by the sun at different times of the day. The sundial is a flat plate with numbers arranged in a circle and a stick in the middle called a gnomon (pronounced *no-*

FAST FACT

Later civilisations like the Greeks calculated the true length of the solar cycle to be closer to 365.25 days. The Romans also made a similar observation. It was from here that the idea of leap year was born. A calendar reform passed by Julius Caesar in 23 BCE added an extra day to the calendar every fourth year to synchronise with the astronomical or seasonal year.

mon). As the sun progresses through the day, the shadow of the stick moves around the plate, landing on the numbers that denote the time of the day.

There is convincing evidence that the oldest known sundial originated in Egypt around 1500 BCE. Until recently, it was thought that such devices did not come into popular use until the Greco–Roman period, at least 1,000 years later. However, in 2013, a group of researchers from the University of Basel, Switzerland, made a significant discovery when clearing the entrance to an ancient tomb in Egypt's Valley of the Kings. The researchers found a flattened piece of limestone, called an ostracon, with a semicircle drawn on it (see image below). The semicircle was divided into 12 distinct sections, presumably representing the hour lines associated with daylight. The sections were evenly spaced, and small dots in the middle of each section allowed time to be measured more precisely. The hole would have held a metal or wooden bolt that cast a shadow that fell on the markings, giving the time of day. This archaeological find is believed to be the world's oldest known sundial.

Telling time seems to have been very important even 3,500 years ago. This small limestone sundial from around 1500 BCE is the oldest portable sundial found so far. It was easy to carry wherever Egyptians travelled.

Interestingly, the limestone sundial was discovered in an area where workers in that time were building graves for the Egyptian pharaohs and nobility. One theory is that the sundial was created by the workers to measure their working hours.

Sundials were further developed by the Greeks, who made various mechanical improvements. You can still see sundials in use today, but they are mostly used for decorative purposes. While sundials are a very inventive way to track the time of day, their accuracy tends to vary depending on their design, how they are positioned relative to the earth's rotational axis, and their geographical location. For example, the shadows on a sundial will cast differently depending on whether it is in the Northern or Southern Hemisphere. Nearly all (if not all) timekeeping devices were developed in the Northern Hemisphere. When sundials were brought to the Southern Hemisphere, it may have been surprising to discover that the gnomon's shadow did not travel the "correct" way around the face of the dial.

A sundial in the Royal Tasmanian Botanical Gardens, Hobart, Australia. If you look closely, you can see that the hour marks run anticlockwise.

ACTIVITY 19:
Resourceful rope stretchers

Go to page 100 to complete the activity and learn about the triangles the ancient Egyptians could make with a very useful tool.

ACTIVITY 20:
Working all the right angles

Go to page 101 to complete the activity and work with the cubit, the length unit that was commonly used in ancient Egypt.

Geometry – all about angles

The origins of right-angled triangle geometry can be traced back to 3000 BCE in ancient Egypt. The ancient Egyptians realised that structures would be more stable if they were built with right-angles at each corner. The Egyptians found a way of using knotted ropes to create square corners. They arranged the rope in the shape of a triangle with sides that measured three knots, four knots and five knots long. The corner where the two shortest sides met would make a right-angle. Rope stretching was not only practical but also

Egyptian rope stretchers had specialised skills much like a modern land surveyor.

The Egyptians

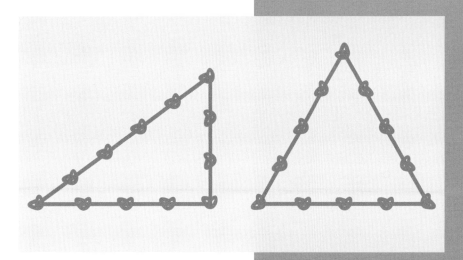

Equally spaced knots in a length of rope were used for linear measurements. If the rope was tied in a continuous loop with a mathematically appropriate number of spaces, it could be used to measure or create special angles. A loop with 12 spaces enabled the stretchers to make right-angled and equilateral triangles.

held ceremonial significance for the ancient Egyptians. Whenever a new sacred building was commissioned, the occasion was attended by pharaohs and other high-ranking officials who personally stretched ropes to define the foundation.

It wasn't until the Greeks came along that a formula was proposed for calculating the dimensions of right-angled triangles. This formula, known as the Pythagorean theorem, is arguably one of the most important advances in the history of mathematics. The theorem is named after the ancient Greek mathematician Pythagoras, who lived about 2,500 years ago. It is thought that Pythagoras didn't actually invent the theorem – he probably learnt it when he was studying in Egypt. So the mathematics we use today to calculate right-angled triangles originated in ancient Egypt and is over 5,000 years old!

The ancient Egyptians would not, of course, have worked with Pythagorean notation. But they clearly had some understanding of the special properties of right-angled triangles and used those properties to build structures that are still standing today.

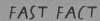

FAST FACT

In geometry, a right-angled triangle is a three-sided figure with one angle equal to 90 degrees. A 90-degree angle is called a *right angle*, giving the right-angled triangle its name. The side of the angle opposite the right angle is called the hypotenuse. The other two shorter sides of the triangle are call legs. The word *hypotenuse* comes from the ancient Greek word meaning "to stretch", which makes sense because the hypotenuse is the longest side of a right-angled triangle!

Pyramid paradoxes

The pyramids of Giza have captivated humanity for more than 4,500 years. They are the last of the Seven Wonders of the Ancient World still standing. Historically, the pyramids testified to the power of the pharaonic religion and state. The three main pyramids were built over the span of three generations for the great pharaohs of Egypt – to house their bodies after death and help them achieve eternal life in the underworld. Each pyramid was made by and for a different pharaoh: the first and largest (known as the Great Pyramid) was built for King Khufu around 2550 BCE, the second was dedicated to King Khafra around 2520 BCE, and the third and smallest pyramid was built in honour of King Menkaure in about 2490 BCE.

Today, the pyramids are famous not only for their royal legacy but also for their size and the amazing skill with which they were built. Many people marvel at how the ancient Egyptians were able to build such huge structures using the technology available 4,000 years ago. Building pyramids

The great pyramids of Giza form a group near the Nile River and just south west of Cairo at the city of Giza. There are many smaller pyramids in the complex, which also includes the famous Sphinx, a human-headed creature with the body of a lion.

of such immense size presented many problems of both organisation and engineering. It has been estimated that more than two million blocks of stone were needed to make the Great Pyramid of Khufu alone. Each block weighed somewhere between 2 and 15 tonnes. These blocks are stacked together into a structure covering the length of two football fields at the base and rising in a perfect pyramidal shape 146 m into the sky. Vast numbers of workers were needed for this task as millions of heavy stone blocks not only needed to be quarried and lifted and stacked to form the pyramid, but also set together with precision to create this shape and ensure stability.

One of the most remarkable things about the great pyramids is how well they have been preserved despite being built so long ago. Before the pyramids were built, the Egyptians built other structures that have not lasted as well. This is because in the other buildings the stones were not cut accurately enough to ensure they could be aligned with no gaps. This allowed moisture to get into the gaps, eventually pushing the stones apart. The Egyptians saw what was happening and realised that if they could construct joints so tight that water couldn't get in, the building would not be destroyed by the water, and it would last a long time. So this is what they did with the pyramids. Another reason the pyramids have lasted so long is that the Egyptians used a special kind of stone similar in strength to granite – this prevented water from being absorbed by the stone and eroding it over time.

Just as impressive as their choice of building material was the way the Egyptians were able to transport the heavy stone blocks and stack them in shape. The answer could lie in a 2018 research project led by the University of Liverpool. A team of researchers discovered a well-preserved ramp at a quarry site near modern-day Luxor (Upper Egypt). The ramp dates to the reign of Pharaoh Khufu and might explain how heavy building stones were transported, lifted and then stacked to form the pyramids. The ramp had a stairway on each side with posts at certain points along the stairs. It is believed that rope would be anchored to the posts and used as a pulley system to haul the stones up the ramp.

Another mysterious feature of the Great Pyramid is the concave design of its outer faces. An aerial photograph taken in 1940 showed that the sides of the Great Pyramid are slightly indented from base to peak down their central lines. This effect divides each of the apparent four sides in half, creating a pyramid that paradoxically has eight outer faces instead of the conventional four! Various theories have been put forward to explain this unique structural property. Some theories point to aesthetic reasons or the effects of natural erosion, while others suggest the faces were intentionally built this way for structural stability. For now, we can only speculate.

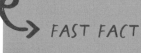

FAST FACT

The concavity of the Great Pyramid is almost invisible to the naked eye. It was only discovered when a pilot flying over it in 1940 at just the right time of day noticed the different shades of the two halves of one "side". His photograph of the concave sides became famous.

Aerial photographs suggest that the pyramids at Giza have eight sides at the base and are therefore slightly concave. As a result, these objects might be called "concave octagonal-based pyramids". Shapes that point inwards (such as arches and dam walls) are stronger than straight shapes, which may be the reason for the concave nature of the pyramids' faces.

Another interesting aspect of the Giza pyramids is the near-perfect alignment of their sides with the four cardinal directions (north, south, east and west). According to one calculation, the pyramids are so precisely aligned as to be no more than one-fifteenth of a degree off true north, true south, true east and true west. As yet, no ancient compass or equivalent surveying device has been discovered to make sense of this extraordinary feat. It seems that the Egyptians used their knowledge of right-angles to position and orient their pyramids in such a precise way. A number of ideas have been put forward by scholars to explain the careful alignment of the pyramids, including the use of constellations and the sun. One theory is that the Egyptians developed their concept of direction from their use of sundials on the fall equinox. On this day, the Egyptians observed that the tip of the sun's shadow ran a near-perfect straight line from east to west. This allowed them to determine the direction of true north by observing the position of the shadow in the middle of the day. Although this is one possibility, the mystery has never been finally resolved. Whether it is a result of pure coincidence or pure genius, it is no wonder the great pyramids have been described as the "most accurately oriented edifices on earth".

ACTIVITY 21:
Working with degrees

Go to page 102 to complete the activity and work with compass directions to explore the regions within and around Egypt.

ACTIVITY 22:
Egyptian mathematics word search

Go to page 103 to complete the activity and review some of the terms related to the mathematics of ancient Egypt.

The Egyptians

Summary

Egypt is an archaeological treasure trove. Much of what we know about this civilisation comes from artefacts found in and around the tombs of pharaohs inside the pyramids.

Numbers

- Devised notation for working with unit fractions
- Used the doubling method as the basis for multiplication
- Used objects from the environment to write different powers of 10

Measurement

- Constructed one of the oldest portable sundials to keep track of time
- Devised a calendar that estimated the length of a solar cycle to be 365 days, which gave rise to the need for leap years

Geometry

- Created right-angled triangles from knotted ropes with sides in the ratio of 3:4:5
- Built pyramids with faces positioned in near-perfect alignment with the four cardinal directions

The Egyptian legacy

The Egyptians used mathematics to solve day-to-day practical needs. The pyramids of Giza are the last of the Seven Wonders of the Ancient World still standing!

Activity 12: *Where in the world is Egypt?*

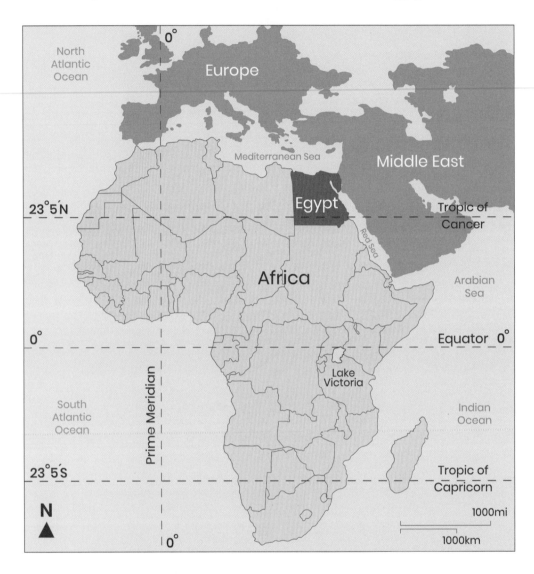

1. What direction do you travel from Egypt to reach:

   a. the equator? _____

   b. the prime meridian? _____

   c. the Atlantic Ocean? _____

   d. Europe? _____

2. Mark the point where the prime meridian and the equator meet with an (X).

   What are the coordinates of this point? (_____ , _____)

3. The Tropic of Cancer passes through Egypt.

    a. What is the latitude of this important line? _____

    b. What solar event occurs on this line once every year?

_____

4. Write the name of a continent that has any part of it in a quadrant of the earth's coordinate system, as listed below. Use the map at right to help. You could have multiple answers for a quadrant.

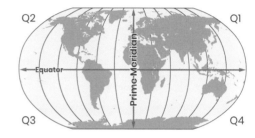

Q2:	Q1:
Q3:	Q4:

5. Use a ruler to estimate each of these distances from the map on the previous page.

    a. The east–west distance across Egypt is about _____ km.

    b. The north–south distance across Egypt is about _____ km.

Activity 13: *The Egyptian number system*

1. Write the value of these Egyptian numbers:

   a. ||| ∩∩ _____

   b. ||||| ∩∩∩ 9 9 _____

   c. 9 9 9 9 || _____

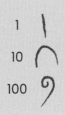

1	|
10	∩
100	9

2. Complete each of these tasks using the number 156.

   a. Write the number using Egyptian symbols.

   _____

   b. Use the same symbols to write 156 another way.

   _____

   c. What is the same about the two ways you wrote the number?
   What else did you notice?

   _____

3. Write the Egyptian number that is one less and the number that is one more.

One less than the number	The number	One more than the number
	|| ∩∩∩ 9	
	||||| ∩	
	||||| 9	

Activity 14: *Decoding Egyptian hieroglyphics*

1. Write the value of these Egyptian numbers.

1	|
10	∩
100	ϱ
1,000	⚘
10,000	(
100,000	𓆐
1,000,000	𓀀

a. | ∩ ∩ ϱ ⚘ _____

b. | | | ∩ ∩ ∩ ∩ ∩ ϱ ϱ ⚘ ⚘ ⚘ (((

_____

c. | | | | ∩ ϱ ϱ ϱ ϱ ϱ ϱ ϱ ϱ ϱ ((𓆐

_____

d. | ϱ ⚘ ⚘ ⚘ ⚘ ⚘ 𓆐 𓆐 𓀀 𓀀 𓀀 𓀀 𓀀 𓀀 𓀀

_____

2. Draw hieroglyphs to show how the Egyptians would record these numbers.

a. 678

_____

b. 7,871

_____

c. 2,353,411

_____

3. What is value of this Egyptian number? How did you work out the value?

(| | ⚘ ⚘ ⚘ ϱ ϱ ⚘ ⚘ ⚘ ⚘ ⚘ ⚘

_____

_____

_____

Activity 15: *Multiplication by doubling*

" The Rhind Papyrus, a work copied by the scribe Ahmes in 1650 BCE, has solutions to many mathematical problems. It is named after a Scottish book collector, Alexander Henry Rhind, who bought it in Egypt in 1858 and bequeathed it to the British Museum. The multiplication method shown here is essentially binary and emulates what computers do today.

Follow these steps to multiply two numbers.

	Multiply 9 and 34	
Write 1 and the greater factor in separate columns.	1 | 34	When we multiply, the two numbers are called factors.
Double the numbers in each sequence. Stop just before the number in the first column exceeds the first factor.	1 | 34 2 | 68 4 | 136 8 | 272	If we double again, the next number is 16 which is greater than 9. So we stop at 8.
Select rows where numbers in column 1 equal the first factor. Add the selected numbers in column 2.	1 | 34 2 | 68 4 | 136 8 | 272	I see that 1 plus 8 is 9. So, I will add 34 plus 272. That means the answer is 306.

1. Complete each of these examples to find the products indicated.

a. 17 × 35 = _____		**b.** 24 × 42 = _____		**c.** 28 × 45 = _____	
1		1		1	
2		2		2	
4		4		4	
8		8		8	
16		16		16	

Activity 16: *Using Egyptian numbers to multiply*

1. Use the Egyptian doubling method with their numerals to answer each of these. Regroup where you can.

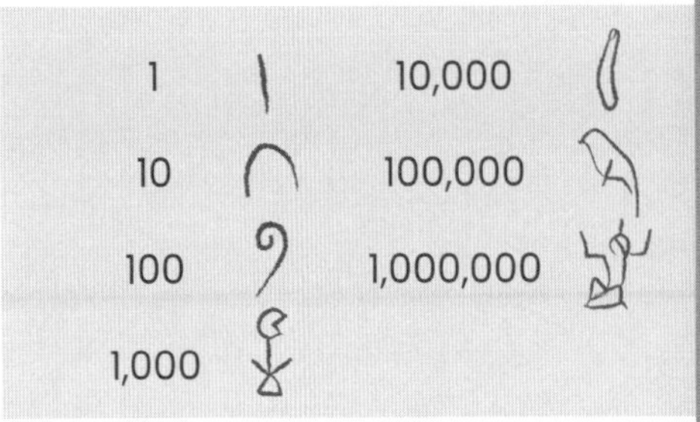

1	$\mid$	10,000	
10	$\cap$	100,000	
100		1,000,000	
1,000			

a. 12 × 52 = _____

$\mid$	$\mid\mid\cap\cap\cap\cap\cap$

b. 17 × 34 = _____

$\mid$	$\mid\mid\mid\mid\cap\cap\cap$

c. 12 × 102 = _____

$\mid$	$\mid\mid$?

d. 21 × 115 = _____

$\mid$	$\mid\mid\mid\mid\mid\cap$?

Activity 17: *An eye for fractions*

" The Egyptian capacity measure for grain was a *hekat*. If the amount of a sale was less than one *hekat*, they used special notation from the Eye of Horus for the fractional parts.

Note: the fraction symbols shown here are for reference. The Egyptians *did not* use these modern symbols.

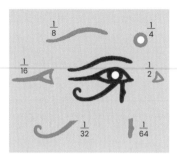

1. What kind of fractions were shown in the Eye of Horus? (Look at the numerators.)

_____

2. What fraction of a *hekat* does each of these represent?

a.	**b.** ⭕	**c.**
d.	**e.**	**f.**

Often, the symbols had to be combined to show multiple fractions.

3. Work out the value of each of these groups of symbols. Write the answer as a single fraction.

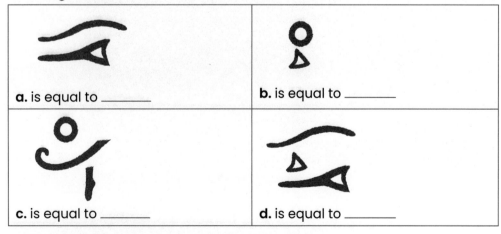

a. is equal to _____

b. is equal to _____

c. is equal to _____

d. is equal to _____

Activity 18: *Greedy algorithms*

" The Egyptians came up with a method to write fractions as the sum of different unit fractions. A unit fraction has 1 in the numerator.

1. Make jumps from 0 to each fraction on the number line. For each example, *every jump* must be the greatest possible unit fraction, for the distance remaining. The denominators of all fractions must be different. The goal is to make the fewest number of jumps possible.

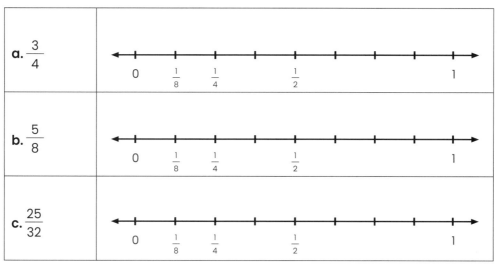

2. Determine how the Egyptians would have written these fractions as the sum of different unit fractions. Use equivalent fractions to help.

Hint: subtract the first unit fraction from the original fraction

a. $\dfrac{3}{5} = \dfrac{1}{2} + \dfrac{1}{?}$	
b. $\dfrac{3}{10} = \dfrac{1}{4} + \dfrac{1}{?}$	
c. $\dfrac{11}{12} = \dfrac{1}{2} + \dfrac{1}{?} + \dfrac{1}{?}$	

Activity 19: *Resourceful rope stretchers*

A specialist group of Egyptians used ropes to measure land and buildings. Many times, the rope was tied in a loop. In this form, it could be stretched tight to make triangles.

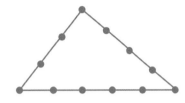

1. Circle all of the names that could be used to describe this triangle.

ACUTE   OBTUSE   EQUILATERAL

SCALENE   ISOSCELES   RIGHT

2.a. Write the lengths of the three sides used to form the triangle above.

 Clue: the sum of any two sides must be greater than the third side.

_____ : _____ : _____

b. What could be some other lengths of sides for triangles that could be made with this chain?

_____ : _____ : _____   _____ : _____ : _____

3. Study the shape shown at right.

a. How many individual links make up the entire chain? _____

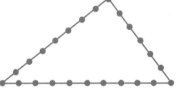

b. Write the lengths of the sides for each type of triangle listed here that can be made using this chain. Be careful to make sure each set of numbers does make a triangle.

The equilateral triangle	
Some isosceles triangles	
Some scalene triangles	

Activity 20: *Working all the right angles*

The Egyptians used the *cubit* for the distance between the knots. The cubit was usually the distance from the Pharaoh's elbow to the end of the longest finger.

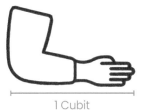

1 Cubit

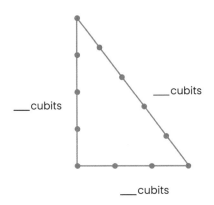

___cubits

___cubits

___cubits

1.a. How many cubits can you count around the perimeter of this triangle?

_____ cubits.

b. Write the length of each side of the triangle.

2.a. Write the length of each side of this triangle.

b. Look at the dimensions for both triangles. What do you notice?

_____

_____

_____

_____

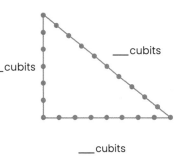

___cubits

___cubits

___cubits

c. Check that both triangles on this page are right-angled triangles. Then mark the square corners on each triangle.

Use the Pythagorean theorem to check if this is a right-angled triangle.

Activity 21: *Working with degrees*

The sides of the Khufu Pyramid are in near-perfect alignment with true north–south. Use this fact to complete these tasks.

1. If the compass bearing of north is zero degrees (0°), write the bearings that fall along each of the solid lines and then each of the dotted lines.

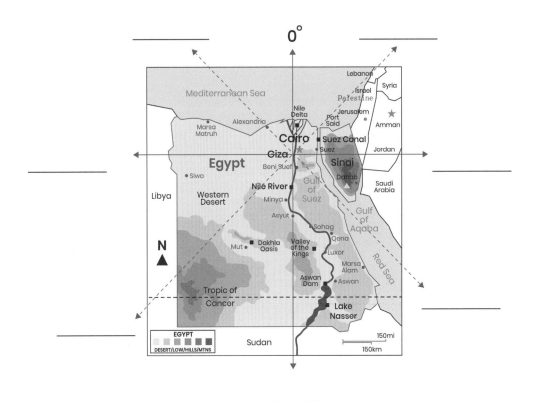

2. Estimate the compass heading for each of these landmarks.

    a. Alexandria: _____ degrees

    b. Jerusalem: _____ degrees

3. Name at least one landmark that intercepts with these compass headings.

    a. 161 degrees: _____

    b. 259 degrees: _____

Activity 22: *Egyptian mathematics word search*

1. Find these words that are hidden in the grid below. All words are written horizontally left to right, vertically top-down, or diagonally top-down (sloping to the right or to the left).

EGYPTIAN   NILE   HEKAT   DENOMINATOR   SIRIUS

PAPYRUS   ADDITIVE   UNIT   SCRIBE   LEAP YEAR

HIEROGLYPHICS   TEN   FRACTIONS   ALGORITHM   PHARAOH

LOTUS   HORUS   NUMERATOR   REMAINDER   ROPE STRETCHERS

C	S	X	P	T	E	O	L	E	F	R	A	C	T	I	O	N	S	K	I
O	R	P	B	C	D	H	D	P	Z	R	S	P	F	Q	M	A	E	L	W
P	B	G	Y	E	O	T	I	H	Q	F	M	N	A	J	V	B	N	P	R
T	U	W	D	G	Z	V	C	E	C	C	A	L	N	I	X	T	J	B	O
B	I	Q	E	P	S	N	Y	M	R	O	G	R	J	X	U	E	C	R	P
Z	Q	H	I	P	N	O	R	J	S	O	U	X	P	Q	H	N	R	M	E
V	E	H	E	K	A	T	W	V	R	A	G	T	P	B	J	R	S	E	S
P	S	C	R	I	B	E	Z	I	V	L	M	L	O	T	U	S	I	O	T
E	F	K	J	K	J	X	T	W	M	T	P	N	Y	C	K	R	H	S	R
S	H	P	Y	T	A	H	D	G	H	G	D	C	L	P	I	Q	Z	X	E
N	O	E	Y	C	M	Y	P	M	E	F	J	P	H	O	H	R	N	I	T
K	V	R	H	C	S	P	A	X	P	R	S	A	E	I	E	I	P	G	C
D	G	J	P	G	W	N	P	M	H	Q	R	E	P	T	L	Z	C	B	H
K	B	J	N	O	T	X	Y	O	U	A	Y	L	J	E	X	Z	Y	S	E
N	D	K	D	R	L	P	R	E	O	Z	U	B	S	C	B	F	F	U	R
G	O	Y	E	C	I	U	U	H	M	W	O	S	G	D	F	K	E	O	S
F	N	L	N	J	S	Q	S	J	P	X	Y	A	N	U	R	K	U	C	C
E	J	U	O	A	Z	I	T	B	K	B	D	R	E	Q	P	N	N	H	L
G	U	Q	M	A	P	Y	R	V	F	L	J	A	D	D	I	T	I	V	E
Y	D	P	I	E	E	N	K	I	C	T	I	Y	E	W	P	V	T	N	A
P	K	M	N	L	R	L	B	U	U	R	N	M	F	O	V	A	E	A	P
T	P	J	A	N	N	A	H	M	E	S	E	H	G	D	S	K	I	L	Y
I	U	R	T	W	Q	Z	T	B	O	K	T	I	Y	S	C	V	N	N	E
A	M	R	O	W	B	E	B	O	K	T	E	S	I	U	E	X	U	B	A
N	J	Q	R	L	L	M	V	U	R	X	R	E	M	A	I	N	D	E	R

CHAPTER THREE

The Indus Valley

When? Where? How?
Who? Why?
I wonder, wonder,
wonder, wonder.

Where did the idea for our modern number system begin? Name the modern-day countries in that location.

Some people think there might be a number system that is better than base 10. What could it be?

What would we do if we did not have a symbol for nothing?

What are different ways to compare two numbers such as 120 and 30 cm?

Two thousand years ago, people had methods to multiply numbers such as 89 by 96 in their heads faster than we can today. How did they do it?

The Indus Valley:
origins of our decimal number system

The Indus Valley river system supported a large civilisation that may have originated before the civilisations in Mesopotamia or even Egypt. But for a long time the contributions this civilisation made to mathematics were contested. In 1817, English mathematician and Oriental scholar Henry Thomas Colebrooke completed the first translations of famous Indian mathematics. However, people in the West were sceptical that the mathematics had been developed completely independently and without influences from Europe and the Greeks. Support for the European origin of mathematics came from the first history of mathematics, a highly influential four-volume series published by German mathematician and historian Moritz Cantor between 1880 and 1908. Arguments regarding the origin of Indus Valley mathematics – was it Indian or Greek? – lasted for over 100 years and influenced the history of mathematics well into the 20th century. For example, in 1907, Florian Cajori described Hindu mathematics in tones that suggested their ideas, while unique, came from the Greeks or possibly the Arabs. Now, with modern technology letting us more accurately date artefacts, the matter is resolved. The work of the Indus Valley mathematicians was completed well before any influence from the Greeks was possible.

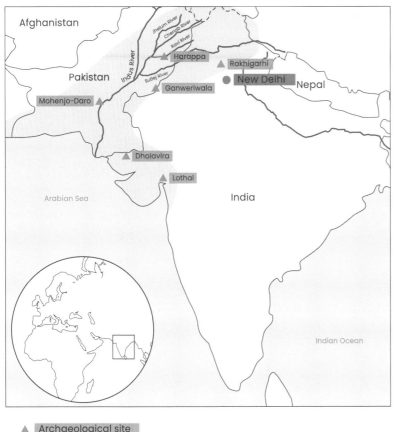

▲ Archaeological site
● Modern city

The Indus civilisation was established about 5,000 years ago at Harappa on the Ravi River, a tributary of the Indus River. It produced some of the first mathematical ideas. About 3,500 years ago, the Harappan culture was changed dramatically by an invasion of pastoral peoples from the north called "Aryans". These were Indo-Aryan from central Asia, different from the Euro-Aryan that was often described as the "Aryan race" – a label adopted by some groups in the early 1900s for a supposedly exclusive and superior race of people, and which is now, of course, no longer in use.

ᔈ
ACTIVITY 23:
The early centres of mathematics

Go to page 130 and complete the mapping activities about other early civilisations near the Indus Valley.

The invasion by the Indo-Aryans was likely facilitated by the invention of the wheel, which allowed people to move more freely. This invasion is usually classified as a friendly invasion as it occurred slowly with ideas from the Harappan culture merging with those of the Aryan. This merger of the two cultures gradually spread further south, expanding throughout the Indus River valley and then along the coast, east and west from the mouth of the river.

The Aryans brought a language that was eventually called Sanskrit. Initially it was a spoken language, but its systematic nature made it highly suitable for scientific discussions, so over time a written form of Sanskrit developed. This language was used to develop, share, and record mathematical ideas for over 2,000 years.

The British discovered the ancient Indus civilisation in 1829, but early excavations were limited to fewer than 100 sites of the approximately 1,000 "cities" that have since been identified. It was not until the 1900s that useful archaeological information was obtained from these sites. The most significant sites are located at Harappa, Mohenjo-Daro on the Indus River, Dholavira and Lothal – the southernmost city of the civilisation along the coast of modern India.

Number – an early place-value system

Each of the six ancient civilisations discussed in this book had a number system. The Babylonians worked with groupings related to 60 while the Maya developed a base-20 system. The Egyptian, Indus Valley, Inca and Chinese cultures all worked with groupings and/ or symbols using a base of 10, like we use today. But only one number system developed in ancient times is still in use today: the Indus Valley number system. From the start, this system relied on place value, and it evolved to eventually use one set of symbols (numerals) that varied in value depending on the place where they were written.

Over 2,000 years ago, the early inhabitants of the Indus Valley developed a set of symbols for collections containing one to nine objects, as shown on the next page. These first sets of symbols were only used by educated individuals in warrior and priest castes, who were known as Brahmins. As a result, these symbols are often called "Brahmi numerals". The numerals for six and seven are close to what we use today. Also, it is relatively easy to see how the horizontal lines for two and three could have easily transitioned to the modern numerals 2 and 3 if the pen was not lifted off the parchment.

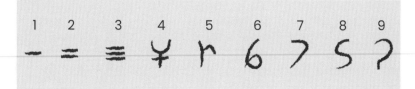

Before the idea of place value was fully established, two different sets of symbols were used for values less than 10 and for groups of 10. The symbols below were used to show groups of 10. Note that a symbol for zero had not yet been established.

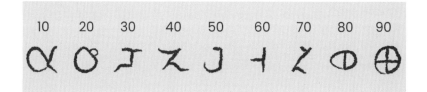

Two-digit numbers were therefore represented by drawing the tens symbol and the ones symbol together, such as the numeral for 47 shown here. This is very similar to the method used by the Greeks when they wrote numbers.

About 1,400 years ago (during the 6th century CE), our modern-day place-value system using one set of symbols for all places became the standard method of writing numbers in the Indus Valley region. Eventually, it became standard across the world. Throughout the transition, the symbols gradually changed to the second set shown in the list on the next page. This set is called "Gwalior numerals" because they were found in a temple at the fort in Gwalior,

The Indus Valley

India, just south of Agra near the Taj Mahal. This set is given special attention because it includes the first record of a zero in India, dating to about 875 CE. There is evidence that the symbol for zero was used in many parts of Asia; for example, a symbol for zero found on a stone in Cambodia has been dated to 683 CE, which is approximately 200 years earlier than the earliest zero found in India. The names for the Gwalior numerals 1 to 9 were *eka*, *dvi*, *tri*, *catur*, *pancha*, *sat*, *sapta*, *asta* and *nava*. *Sunya* was the name for zero.

Over time, and with influence from other cultures and mathematicians, the shapes of the numerals gradually changed to the modern form we use today. Study the list below and see how the early numerals evolved into the set we use today.

Brahmi, 1st century CE

Indian (Gwalior), 9th century

Sanskrit Devanagari,
Indian, c. 11th century

East Arabic, c. 11th century

West Arabic (Gobar),
c. 11th century

15th century

16th century (Dürer)

ACTIVITY 24:
Reading and writing Indus numbers

Go to page 131 to practise using the first numerals that formed part of our modern number system.

Number – working with large numbers

The earliest civilisation in ancient India was the Vedic civilisation. It is named after the Vedas, the oldest Hindu texts. The Vedas began as an oral tradition that was passed down from one generation to another. The Vedic civilisation flourished along the Saraswati River, just to the east of the Indus River in the Indus Valley region that now comprises the modern Indian states of Haryāna and Punjab.

The Vedas provide many examples of large numbers that seemed to focus on groupings that we would today call the "powers of 10". Indian-born mathematician and author George Gheverghese Joseph wrote that a significant and unique contribution from this culture was the use of different names for every power of 10 from 10^1 to 10^{62}. The Hindus were very confident with large numbers, which contrasts with the Greeks who could not work with numbers beyond 10^4 (10,000). As Joseph states, "the importance of these number-names in the evolution of the decimal place-value notation cannot be exaggerated".[2] The Hindus of the Indus

[2] GG Joseph (2010) *The crest of the peacock: non-European roots of mathematics*, 3rd edn, Princeton University Press, New Jersey.

The Indus Valley

Valley had both practical, literary and religious reasons for devising names for such large numbers.

One reason for needing to understand such large numbers was the Hindus' fascination with the vastness of time and space. This seemed to create a fascination for large numbers. While they did not have symbols for these numbers, the Hindus described the vastness of the universe using names for powers of 10. For example, one measurement unit used in ancient India for distance was the *yojana*. The length of this unit varied, but generally was 12 to 15 km. Using this large number, Indus Valley mathematicians were able to estimate the distance from the earth to the sun as 12 million *yojanas* (147 million km). The distance to the sun is 151.28 million km, so 12 million *yojanas* is a very good estimate!

As well as estimating the distance of the sun, the Vedas describe the use of large numbers to record information such as the size of armies. Rama, a famous hero in one of the Vedas stories, was said to have under his command $10^{10} + 10^{14} + 10^{20} + 10^{24} + 10^{30} + 10^{34} + 10^{40} + 10^{44} + 10^{52} + 10^{57} + 10^{62} + 5$ men! Note that this is modern-day notation; in ancient India, words would have been used for these numbers.

Algebra – multiplying mentally

The period from about 200 BCE to 400 CE is usually described as the era of Jaina mathematics. Jainism was a belief system that coexisted with Hinduism and Buddhism, and each religion influenced the other. This period was a time of dynamic development in mathematics, including considerable development of rules in mathematical operations and decimal place notation; algebra including simple, simultaneous and quadratic equations; square roots; and how to deal with unknowns.

Evidence suggests that Indus Valley mathematicians used their knowledge of place value to multiply mentally. While some people question the extent of their capability, the more significant point is that this civilisation began to explore methods that could be used to find unknown values. One example involves finding products of two numbers that are both close to 100. The thinking is based on the geometry of a rectangle, as shown on the next page. This strategy for multiplying large numbers is used in classrooms today.

The example shown here involves multiplying 112 by 106. Begin by drawing a rectangle with dimensions equal to the two numbers.

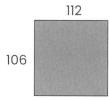

The next step is to break the rectangle into parts then multiply the dimensions of each smaller rectangle.

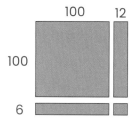

The 100 by 100 square is a power of 10 that the Indus Valley mathematicians could confidently calculate because of their place-value system. The two medium-size rectangles were also relatively easy for the Indus Valley mathematicians to calculate because one dimension is 100. Finally, the small rectangle has dimensions that are easy to calculate quickly because the factors are ones or tens.

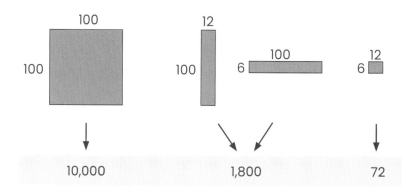

The final step involves adding the three values. This was usually easy because the Indus Valley mathematicians would align the numbers according to place value, as shown here.

10,000
1,800
72

The original numbers used in the method above are both a little more than 100, so it is easy to split the numbers and work with easy factors, such as 100. The ability to multiply numbers this way is another example of the mathematical power of a place-value number system.

When the Indus mathematicians multiplied two numbers that were a little less than 100, they made one of the factors in each number 100, multiplied the dimensions of each rectangle like normal and then subtracted from the total what they added to get to 100. It is interesting to note that this process could have involved working with negative numbers, but we have not yet found evidence of the exact steps that were used.

ACTIVITY 25:
Multiply mentally with algebra methods

Go to page 132 to work through the steps to multiply numbers close to 100 using the Indus Valley methods.

Measurement – weights

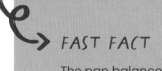

FAST FACT

The pan balance was used for centuries. The shape of the device closely matches the star sign Libra. The Romans called the unit of weight "the libra". The abbreviation was "lb" which was retained when the pound became popular.

In the Indus Valley, the attribute of mass (weight) seemed to attract even more attention than length in terms of mathematical detail. This might be because mass was used in commercial transactions involving grain or even more precious commodities, such as gold. If a precious mineral was involved, both the buyer and seller would want to be sure the measurements were exact, so accurate mathematics was required.

A pan balance is a measuring device that clearly shows the effect of gravity when finding the mass of an object. When the pans are level across, the pull of gravity is the same on both sides. In this way it is possible to find a value for the object with the unknown mass by adding the known values of the mass pieces. The pan balance shown here does not have a pillar or base, which made it easy to transport and carry from one city to the next.

The term *binary* is used for a base-2 system because only two digits, 0 and 1, are needed. On the pan balance, each mass piece is either *on* (1) or *off* (0) the pan being used to find the weight of the object. This is like the on/off states (1/0) used in computer systems today. So the terms *binary* and *base-2* can be used interchangeably.

ACTIVITY 26:
Working with weights

Go to page 133 to analyse a measurement system used long ago. Those ancient methods were binary like the mathematics a computer uses today.

Findings from archaeological sites throughout the region show that pan-balance weighing devices were used to measure mass. Many stone mass pieces have also been found. These would have been placed in one pan to measure the weight of the commodity, as shown below. Interestingly, the units of weight follow both a binary (base-2) and a decimal (base-10) system for the mass pieces and are relatively small in size and weight. From the small size of the mass pieces, we can infer that many of the commodities weighed would have been precious.

Nearly all the mass pieces, like those shown below, were graduated sets of roughly cubical shapes. An analysis of the mass pieces found at three different locations in the Indus Valley (Harappa, Lothal and Mohenjo-Daro) shows that the weights have ratios that (with the exception of the very last value) follow a binary sequence (1:2:4:8:16:32:60). In this case, the base unit has a modern-day mass equivalent of a little less than one gram (the accepted mass at multiple sites is 0.89 g). Sometimes the sequence was a combination of binary and decimal, such as 1:2:5:10:20:50:100:200:500. Interestingly, this sequence is similar to the value of coins and notes used in many countries today!

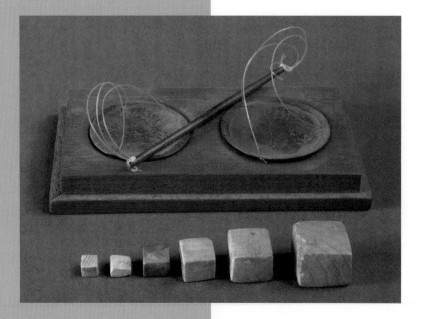

A consistent-looking and regularly increasing sequence of objects made it easy to identify the values for the weights or mass pieces. From a practical point of view, these nearly cube-shaped stone pieces were easy for a merchant to pack and carry for long journeys.

Measurement – building with perfect bricks

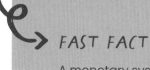

FAST FACT

A monetary system based on a ratio of 1:2:5:10:20:50:100:200:500 with coin and note denominations is the most efficient for paying exact amounts. For example, any amount from $1 to $1,110 can be paid by carrying 12 coins and notes – one of each value in the ratio above plus an extra $2, $20 and $200.

This is how we would pay $404. Note that without $2, $20 or $200 denominations, we would need eight notes.

$200	$200
$2	$2

One of the first cities in the Indus Valley to be excavated by the British was Harappa, a city in the northern part of modern-day Pakistan. As the British were excavating the city, they observed that the bricks possessed interesting mathematical relationships. Unfortunately, they then used the bricks to build the first railroads in the area and so destroyed key archaeological and mathematical data. As the picture here attests, the bricks had lasted for thousands of years and were still strong enough to act as the base for the British-built railways.

Some of the best remaining examples of the Indus bricks can be found in the city of Mohenjo-Daro. This is one of the five largest Harappan cities that formed the ancient Indus civilisation located in modern-day Pakistan. The picture shows how the bricks could be laid to create rooms and walkways with very accurately formed right-angles.

ACTIVITY 27:
Perfect bricks

Go to page 134 to explore more about the special properties of the Harappan bricks.

These bricks are often called "Harappan bricks" after the city where the British first found them. However, most cities in the Indus region also used these bricks, and they can be found everywhere in the Indus Valley and along the coast. The bricks were basic. They were handmade – not cast in moulds – and made of mud that was then baked in the sun. The type of soil in that region helped ensure that the bricks would last a long time.

As the British worked with the bricks to build the first railroads, it soon became apparent that the bricks possessed mathematical relationships that made them ideal for constructing buildings. The bricks were made in varying sizes, but the length to width to height ratio was always 4:2:1. Many of the brick walls throughout the civilisation that still stand today were built thousands of years ago with the layers of bricks placed in alternate directions, which illustrates the 2:1 length to width ratio.

These walls located in Dholavira were built about 4,500 years ago and have lasted through a range of severe weather conditions as well as earthquakes that destroyed other structures built by the civilisation. It is in the south-western part of India and lies on the Tropic of Cancer, an ideal position from which to make solar observations on the Northern Hemisphere summer solstice. One building – the bailey structure – has been identified for this purpose. The image shows part of the reservoir for the city; storing water was necessary in the heat.

The fact that the ratio of the lengths of the different sides of each brick remained constant throughout the geographical and chronological history of the Indus Valley civilisation is very impressive given that the bricks were handmade! Today, this ratio is still considered to be the best relationship for the dimensions of bricks.

More about length

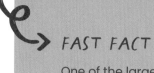

FAST FACT

One of the largest and most impressive cities in the Indus Valley is Dholavira near the southern coast. In the wet season, it was surrounded by water and could be affected by tsunamis, so the people built massive walls to protect the city. According to research conducted by British archaeologist Sir Mortimer Wheeler, it is the only major city of the ancient Indus civilisation built with stone, not bricks, as the mud bricks would have dissolved with the onslaught of water. Dholavira, with its unique architecture, is now a UNESCO World Heritage site.

It is difficult to imagine that the Harrapan bricks, with such consistent ratios of their dimensions, could be made just by hand without a measurement tool. So it is not surprising that excavations have uncovered ruler-like measurement tools. The tools are divided into units that are 3.35 cm (1.32 inches) long and almost always highly accurate. This length, sometimes known as the "Indus inch", has been in use for at least two millennia. Like modern-day rulers, individual units are grouped in tens to reflect the decimal structure of the number system. Ten of these units equal 33.5 cm (13.2 inches). Research suggests that a typical building brick could have been about 10 cm thick (equivalent to 3 Indus inches). Using the ratios described in the previous section, the width and length of the brick would have been respectively 20 cm and 40 cm.

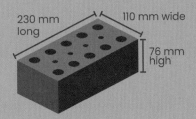

230 mm long

110 mm wide

76 mm high

The ancient Indus civilisation had an intricate measurement system with many different and interrelated units. In addition to the shorter units that related to the Indus inch described above, they had a *pada* (which translates directly as foot), *hasta* (the cubit or distance from the elbow to the end of the longest finger) and *purusha* (the height of a person). There were also longer units of length, such as the *yojana* mentioned earlier. The mathematical relationship between each of these and many other units used in the Indus Valley is not clear.

Measurement units for length were required for various types of constructions. One example is the Vedic altar design shown below. This altar had to be carefully built with the exact dimensions described in *padas*. Note that this design was constructed flat on the ground and used as a fire altar. Naturally, these altars did not survive, so our information about the altars comes from recent translations of old texts.

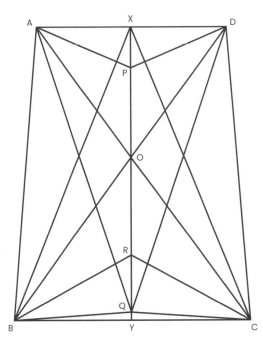

AX	=	XD	=	12 *padas*
BY	=	YC	=	15 *padas*
		XP	=	5 *padas*
		PR	=	23 *padas*
		RQ	=	7 *padas*
		QY	=	1 *padas*
		XY	=	36 *padas*

Geometry

The various shapes and key lengths used to create the Vedic altar reveal some very interesting mathematics that seems to have been intentionally embedded in the design. For example, the diagram is highly symmetrical with the outline forming an isosceles trapezium. There are many triangles and concave quadrilaterals drawn within the trapezium that are also symmetrical. Some of the triangles have a right-angle, for example at points X and Y. From either of these two points, it is possible to identify at least 14 right-angled triangles, such as $\triangle AXP$ (the triangle formed from the points A, X and P) and $\triangle AXO$. Some of the right-angled triangles have side lengths that are easy to calculate and therefore confirm they are a right-angled triangle. The use of right-angled triangles in this altar indicates that there was some knowledge of the properties of right-angled triangles over 2,500 years ago in this region of the world.

Knowledge about right-angled triangles is indirectly supported by verses in the *Sulbasutra* attributed to Baudhayana, sometimes called the Indian Pythagoras.

ACTIVITY 28:
A triangular-based altar plan

Go to page 135 to further investigate the shapes in the fire altar and explore early uses of the side relationship of right-angled triangles. This activity does not require a calculator, but you may use one if you want to be more accurate with your calculations.

Personal information about Baudhayana is scant, but he was probably not a mathematician. His interest was religious, and he used mathematics to plan altars. The important mathematics here involved right-angled triangles and related rectangles. The *Sulbasutra* states that the diagonal of certain rectangles, when squared (shown in the first picture below) produces an area that is the same as the area produced by adding the squared length and squared width of the same rectangle (shown in the second picture below). In the example below, the area of the square yellow region in the first picture is the same as the sum of the area of the two rectangular yellow regions in the second picture.

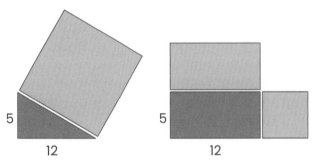

It then gives the lengths and widths of some of these rectangles as 3 and 4, 5 and 12, 7 and 24. This suggests the Vedic mathematicians knew the relationship between the length of the diagonal and the two side dimensions of certain rectangles. This is the relationship that the Pythagorean theorem uses for finding the hypotenuse of a right-angled triangle!

There is no evidence that this understanding of right-angled triangles was influenced from outside sources. So it seems that three different civilisations – Babylonian, Egyptian and Indus – each individually developed understandings of the relationship between the lengths of sides in right-angled triangles. It just goes to show that mathematics is a very important part of human life and civilisation!

As the description and dimensions of the Vedic altar suggest, many constructions in the early Indus civilisation were completed for religious purposes. The religious aspect of these constructions imposed specific rules that had to be followed to achieve correct and religiously worthy results. These specific rules required mathematics. Like other early civilisations, the Indus mathematicians found the circle to be the easiest shape to draw accurately. Interestingly, it also provided guidance to create perfect shapes such as squares. Below is the sequence described in *Sulbasutra* for constructing a square.

Construct an east–west line.

Draw a circle with the midpoint on the line.

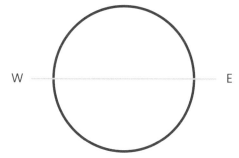

Draw arcs (longer than the radius of the circle) to find the perpendicular bisector and therefore create a north–south line.

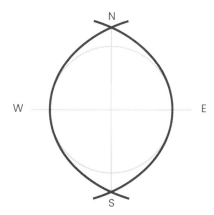

ACTIVITY 29:
Using circles to draw a square

Go to page 136 to further investigate the shapes in the fire altar and explore how circles were used to outline altars with straight sides. This activity requires a straight edge, but you don't need a ruler.

ACTIVITY 30:
Indus civilisation word match

Go to page 137 to complete the Indus Valley word-match activity.

Draw two circles with a radius equal to the diameter of the first circle and centres placed respectively to the east and west (These larger circles do not seem to be vital for constructing the square. They would act as a guide for building the larger altar.)

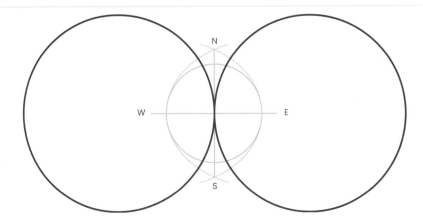

Draw four circles that are congruent to the first circle with centres at the east, west, north and south intersecting points on that circle.

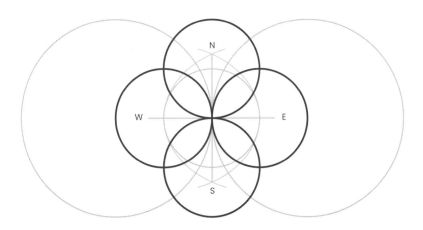

Join the connecting points of the four circles to make a square. The sides of the square could then be extended in one or both directions to make rectangles that helped to outline altars.

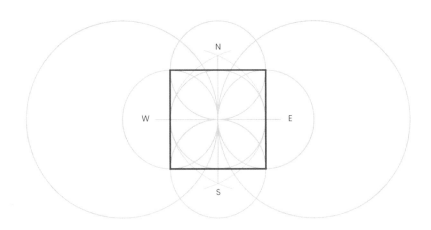

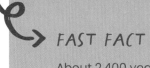

Removing some of the initial guiding lines used to create the square leaves a striking picture!

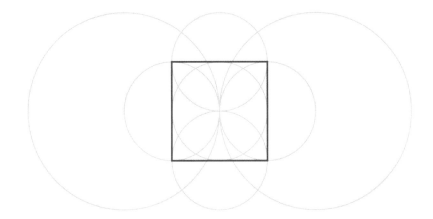

Summary

Indus Valley is named after the large river system near the border of Pakistan and India. It was settled about 5,000 years ago and later influenced by peoples from the north. The mathematics developed in the Indus Valley is the basis of our modern number system.

Numbers

- Used a base-10 decimal system with place value
- Developed a zero about 1,500 years ago (enabled by the place-value system)
- Loved to work with very large numbers and could calculate using methods that are similar to modern-day algebra

Geometry

- Used geometrically precise layout of altars
- Used the relationship between side lengths in right-angled triangles to make precise isosceles trapeziums and rectangles
- Used circles to draw exact square shapes

Measurement

- Had a highly developed system of measuring mass that was used to measure the weight of precious goods
- Always used the same ratio of length units for construction; bricks, in particular, always had the same ratio of the lengths of their sides

Algebra

- Multiplied numbers close to 100 using early ideas related to variables and the ability to generalise steps to "remove" factors from the products

The Indus Valley legacy

By far, the most important legacy of the Indus Valley civilisation is the development of a place-value number system. It is likely that one of the first uses of zero was found in this region.

Activity 23: *The early centres of mathematics*

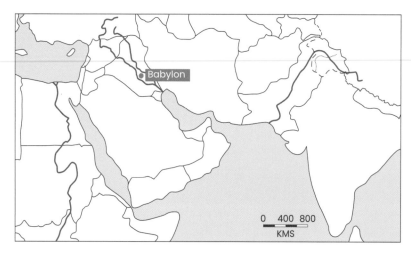

Iraq is the modern country that now includes much of the region that was called Babylon.

1. On the map above, draw a rough circle around each of the three river systems that supported ancient cultures that developed early mathematics. Then write the names of these modern-day countries in their approximate locations.

a. Iraq   **b.** Egypt   **c.** India

2. Use a ruler to estimate the straight-line distance between the following pairs of river mouths.

a. The Tigris and Nile river mouths are about _____ km apart.

b. The Tigris and Indus river mouths are about _____ km apart.

c. The Nile and Indus river mouths are about _____ km apart.

3. The Indus River is located mostly in modern-day Pakistan. Write the name Pakistan in its approximate location on the map.

Egypt	95.7
India	1,324.2
Iraq	39.3
Pakistan	193.2

The Economist Pocket Figures, 2019

4. The populations (in millions) of the countries in the map above are given in this table. Use a number to complete each of these sentences.

a. India's population is about _____ times the population of Iraq.

b. Egypt's population is about _____ the population of Pakistan.

Activity 24: *Reading and writing Indus numbers*[3]

Before developing the place-value system, the first inhabitants of the Indus Valley had different sets of symbols for ones, tens, hundreds and thousands.

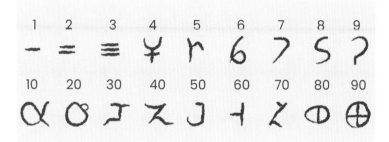

1. Write the value of these numerals.

a. ⟨≡	b. ⟨Ⅎ Ⅎ	c. Ɉ ʔ

2. How would these numbers have been written early in the history of the Indus Valley?

a. 28	**b.** 62	**c.** 36

3. With true place value, one set of symbols would be used for each place. Use the first row of symbols above to write these numbers using place value.

a. 28	**b.** 62	**c.** 36

4. What vital digit is missing in the first set of numbers? _____

3. A small circle ○ was used for the Indus zero. Use it and the numerals in the first set above to write each of these numbers

a. 30	**b.** 106	**c.** 160

³ JBurnett (2005) Sights, sounds and symbols, ORIGO Education, Brisbane.

Activity 25: *Multiply mentally with algebra methods*

1. Follow these steps to multiply each example using the Indus methods.

a. Each dimension above has been "broken" into two parts – one hundred plus some ones. Write the missing dimensions.

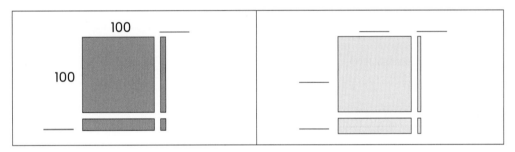

b. Write the areas for the three types of rectangles. Then write the total.

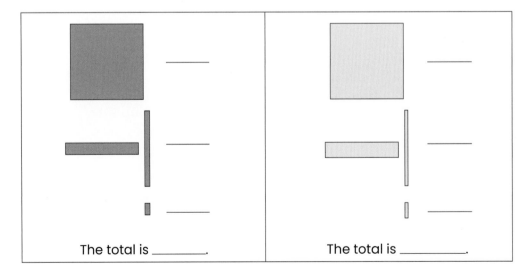

The total is _____.   The total is _____.

Activity 26: *Working with weights*

The Indus mathematicians used stone cubes to work with weights. Each stone was double the mass of the piece just before it. If the sequence begins with 1, the values are the base for binary numbers.

1. Imagine that the first piece in the set of weights below has a mass of 1 kilogram (kg). Write the masses on each of the other pieces.

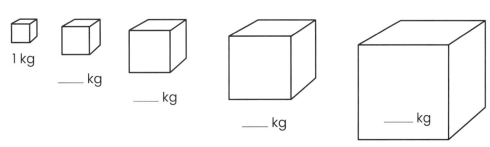

1 kg

_____ kg

_____ kg

_____ kg

_____ kg

2. A binary system of mass requires fewer pieces than a decimal system. List the binary and then the decimal pieces you would use for each of the items below.

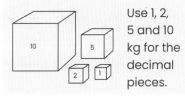

Use 1, 2, 5 and 10 kg for the decimal pieces.

	Binary mass pieces	Decimal mass pieces
a. Gold – 9 kg	How many pieces? _____	How many pieces? _____
b. Dates – 18 kg	How many pieces? _____	How many pieces? _____
c. Wheat – 24 kg	How many pieces? _____	How many pieces? _____

Activity 27: *Perfect bricks*

1. The three dimensions of this brick are shown. Write the three ratios below, in simplest form.

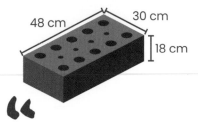

48 cm   30 cm   18 cm

    a. length to width or _____ : _____

    b. length to height or _____ : _____

    c. width to height or _____ : _____

Remember that units are not needed when you write ratios.

2. Indus bricks were constructed using a L to W and W to H ratio of 2:1. Work out the unknown measurements of these Indus bricks.

a.	**b.** 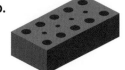	**c.**
length = 60 cm	length = _____	length = _____
width = _____	width = 52 cm	width = _____
height = _____	height = _____	height = 20 cm

3. Calculate the area of the different-sized faces on the brick at the top of the page.

    a. The length by width area is _____ cm².

    b. The length by height area is _____ cm².

    c. The width by height area is _____ cm².

4. Use the answers in Question 3.

    a. Find the ratio of the length by height area to the width by height area, and then write it in its simplest form.

    _____

    b. What do you notice? Look at your answers in Question 1.

    _____

Activity 28: *A triangular-based altar plan*

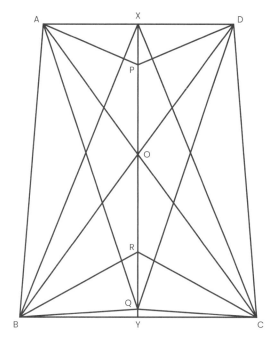

The dimensions of the Vedic altars were very precise.

AX	=	XD	=	12 *padas*
BY	=	YC	=	15 *padas*
		XP	=	5 *padas*
		PR	=	23 *padas*
		RQ	=	7 *padas*
		QY	=	1 *padas*
		XY	=	36 *padas*

1. What special name can you give to angle AXP? _____

2. Write the lengths of the two sides AX and XP. Then calculate the length of AP.

    a. AX is _____ *padas* long.

    b. XP is _____ *padas* long.

    c. AP is _____ *padas* long.

3. What shape is ABCD? _____

4. Write these lengths for shape ABCD.

    a. Base AD = _____ *padas*.

    b. Base BC = _____ *padas*.

    c. Height of ABCD = _____ *padas*.

5. Calculate the area of the entire alter shape (ABCD).

The area of the altar is _____ square *padas*.

Hint: the formula to work out the area of a trapezium is

$$A = \frac{b_1 + b_2}{2} \times h$$

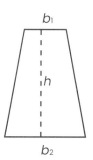

Activity 29: *Using circles to draw a square*

1. If the radius of each smaller circle below is 2 metres (m), what are the dimensions of the square? _____ × _____

2. Extend the two horizontal sides of the square left and right to make a wide rectangle. The vertical sides should touch the circumference of each large circle.

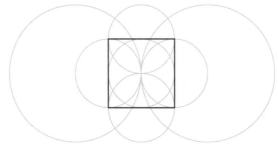

What are the dimensions of this rectangle? _____ × _____

3. Extend the two vertical sides of the square up and down to make a tall rectangle. The horizontal sides should touch the circumference of each small circle.

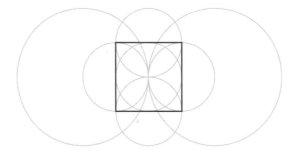

What are the dimensions of this rectangle? _____ × _____

Activity 30: *Indus civilisation word match*

1. Draw a line from the statement to the single word that matches.

The river that was the basis for the civilisation.	The country where the Indus River was located before 1949.	The pattern of pieces used with a pan balance.
A city that is sometimes the name of the civilisation.	A unique numeral used by this civilisation.	The country where the river is located today.

binary	**decimal**	**Harappa**
Hindu	**India**	**Indus**
length	**mass**	**Pakistan**
ratio	*yojana*	**zero**

The base of the civilisation's number system.	An attribute that is measured with a ruler.	A unit used to measure distance.
An attribute that is measured with a pan balance.	One of the religions practised in this region.	A comparison involving the same attribute.

CHAPTER FOUR

Mesoamerica

When? Where? How?
Who? Why?
I wonder, wonder,
wonder, wonder.

What remains of the Maya and Aztec empires could you see today?

When in the year does the silhouette of a serpent famously appear on the walls of the El Castillo pyramid?

Why did the ancient Maya think that having a symbol to represent "nothing" could somehow mean everything?

What type of symmetry did the Aztecs incorporate into their designs?

Why did historians fear the world would suddenly end on 21 December 2012?

Mesoamerica: *the need for nothing*

The historic region of Mesoamerica comprises all or part of the modern-day countries of Costa Rica, Nicaragua, Honduras, El Salvador, Guatemala, Belize and Mexico. The peoples who inhabited these areas shared a number of cultural traits including a complex pantheon of deities, distinctive architectural styles, a 260-day calendar, trade, dress and food (especially a reliance on maize, beans and squash).

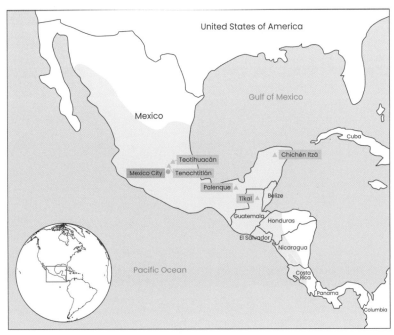

▲ Archaeological site
● Modern city

For thousands of years, Mesoamerica was populated by groups such as the Olmec, Zapotec, Maya, Toltec and Aztec peoples. The geography of Mesoamerica is incredibly diverse – it includes humid tropical areas, dry deserts, high mountains and low coastal plains. An anthropologist named Paul Kirchhoff first used the term *Mesoamerica* (*meso* is Greek for "middle" or "intermediate") in 1943 to designate these geographical areas as having shared cultural traits prior to the invasion of the Europeans in the early 16th century. Many historical artefacts from this part of the world, including pyramids and stone engravings, have survived until the present day. In particular, the pyramids continue to attract visitors from all around the world and reveal important facts about how the Mesoamerican cultures functioned. This chapter focuses mainly on the Maya and Aztec civilisations in which mathematics formed a large part of everyday life. It is important to bear in mind that the Maya and Aztecs developed independently from other cultures such as the Inca. While those three cultures existed at more or less the same time, the Inca developed in a different part of the Americas, and their uses of mathematics were vastly different. We explore the Inca in the next chapter.

ACTIVITY 31:
Mapping the Mesoamericans

Go to page 174 and complete the activity.

The Maya civilisation

ᴹ ACTIVITY 32:
Magnificent Maya

To learn more about the iconic El Castillo pyramid, go to page 175 and complete the activity.

The Maya occupied a region that stretched from the eastern edge of Honduras and northern tip of El Salvador to the northernmost reaches of the Yucatán Peninsula, Mexico. The Maya civilisation reached its peak about 1,800 years ago when their society was ruled by a priestly class that oversaw all aspects of civil and ceremonial life. This period, known as the Classic Era, brought astonishing advances in architecture, astronomy and mathematics. One significant achievement was the El Castillo step pyramid at Chichén Itzá, built sometime between the 8th and 12th centuries CE. Another milestone was the development of a sophisticated number system – one of the first in documented history to make reference to a zero value.

Mesoamerica

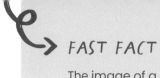

The image of a serpent appears on the walls of the El Castillo pyramid on both yearly equinoxes. Equinoxes are the two days in a year when the hours of night and day are almost equal. These occur when the sun is closest to the equator (about 21 March and 23 September). The serpent head illusion represents the serpent god Kukulcán, in whose honour the pyramid was built.

El Castillo was likely used to track the year. The pyramid has four sides, which could correspond to the seasons. It is located at latitude 20.6 degrees north, so its proximity to the Tropic of Cancer made it almost ideal for tracking the position of the sun at the equinoxes and solstices.

The Maya number system

T he Maya number system was highly advanced for its time, relying on just three symbols to write large numbers and perform basic arithmetic. A dot was used to represent a value of one, a line represented a value of five, and an empty shell represented zero.

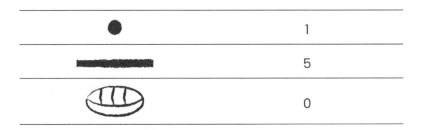

●	1
▬▬▬	5
🥢	0

It is thought that these symbols were created based on items that the Maya people might have first used to count, such as pebbles, sticks and shells. One distinctive feature of this number system is that the Maya wrote numbers in a vertical arrangement, starting with the lowest values at the bottom and working up. Unlike the Egyptians and Aztecs, the position where the Maya wrote their numbers was important: the value of a symbol was dictated by its position in the vertical arrangement rather than how many times it was used. Like we do in our modern-day number system, whenever the Maya wanted to write a number that exceeded the limit in a particular place, they would have to move to the next place value. In their case, the greater place value was positioned above the lesser place value, not to the left like it is in our number system.

While we use a base-10 number system, the Maya used a base-20 or "vigesimal" number system, so each place value increased vertically by a power of 20. We are not certain why the Maya selected 20 as their base value, but it was most likely because young children were taught to count using their fingers and toes.

Maya notation for 2122

400s	▬▬▬▬▬
20s	● ▬▬▬▬▬
1s	● ●

Unlike most other early civilisations, the Maya had two different numeral systems – one was for the common people and the other for the priestly class. The Maya priests considered time to be sacred, and the number system used by the priestly class was mostly for astrological calculations like counting days or other units of time. It could also be used for anticipating sacred events, like a ritual celebration to honour one of the many Mesoamerican deities. The priestly number system was similar to the ordinary number system, except that the third place was counted in units of

360 (18 × 20) instead of 400 (20 × 20). While we don't know why the Maya priests chose a grouping of 360 units in the third place, it was most likely influenced by their perception of time. The Maya solar calendar was based on a 360-day cycle, which was divided into 18 months with 20 days in each. The Maya priests probably understood the benefit of having a mixed-base system to keep track of time, especially the years, over significantly long periods. One example of how this system worked is reflected in a stone inscription from Tres Zapotes, an archaeological site from the Olmec Mesoamerican civilisation. From top to bottom, the numerals on the stone read: 7.16.6.16.18 (see below). The total value of these numbers is the age of the stone according to the Maya calendar. It is expressed as the total number of days that have passed since a date the Maya priests identified as the mythical creation date of the universe.

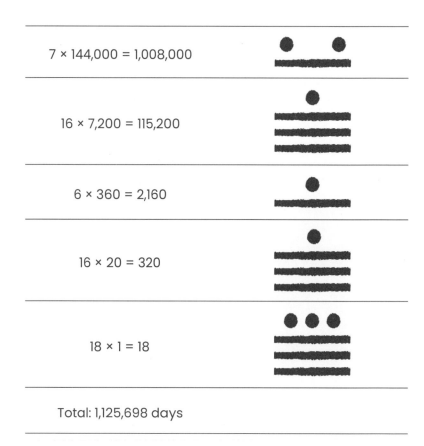

7 × 144,000 = 1,008,000

16 × 7,200 = 115,200

6 × 360 = 2,160

16 × 20 = 320

18 × 1 = 18

Total: 1,125,698 days

The way the Maya grouped their units of time is similar to the way units of time are organised in the modern calendar. The Maya used the *tun* as a benchmark for writing larger units of time like the *katun*, *baktun* and *piktun*. This is similar to the way we use the year as the benchmark for larger units of time like decades, centuries, millennia and so forth. The only difference is that the Maya units of time followed a base-20 progression, whereas the units in the modern calendar follow a base-10 progression. The table below shows how units of time in the Maya calendar compare to those in the modern-day system. In both cases, the underlying base system influences the length (duration) of successively larger units of time.

	Modern calendar	Maya calendar
One year	365 days	365 days
One year	12 "months" of 30 days, plus 5 days or $12 \times 30 + 5$	18 "months" of 20 days, plus 5 days or $18 \times 20 + 5$
One year × base	One decade 10^1 years	One *katun* 20^1 years
One year × base^2	One century 10^2 years	One *baktun* 20^2 years
One year × base^3	One millennium 10^3 years	One *piktun* 20^3 years

Having two different number systems was interesting and unique to the Maya civilisation, but it also created some ambiguity. It meant that one number form could be read in two different ways. It all depended on whether the number was read as an *ordinary* number or a *calendar* number. Ordinary numbers would have been best suited to tasks like commerce, trade and day-to-day record keeping,

ACTIVITY 33:
Cracking the Dresden Codex

The Dresden Codex is one of only a few surviving books of astrological observations written by the Maya about 1,000 years ago. It was given to the King of Spain after the Europeans invaded the Americas and then purchased by the library in Dresden, Germany, in 1744 CE. Go to page 176 and complete the activity.

while calendar numbers would have been used by priests to record important sacred events and prophesise about the future. In each case, the correct way to read a number would probably have been inferred from the context. An example of how one number could be read two different ways is shown below.

ACTIVITY 34:
One number, two meanings

Go to page 177 and complete the activity.

Common people		High priests
10 × 400		10 × 360
7 × 20		7 × 20
13 × 1		13 × 1
= 4,153		= 3,753

When nothing means everything

The origin of the number zero is one of the most contentious topics ever debated by mathematicians. It is beyond the scope of the Mathema series to pass critical judgement on this issue, but it is worth providing some background context. Many ancient civilisations were advanced enough to understand the mathematical concept of "nothingness". For example, the Chinese used counting boards with empty spaces, while the Inca left gaps to indicate no value on the knotted cords called "quipus" that they used to show numbers. However, only a handful of cultures in the history of mathematics are thought to have created a stand-alone symbol for zero. The Babylonians used a double-wedge marker, while the Maya came up with their version of an empty shell. In India, the symbol used to represent zero was a small black dot.

Timeline for zero or nothing as a number

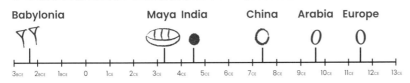

ACTIVITY 35:
A place for zeros

To learn more about the role of zero in the Maya number system, check out page 178.

Whichever ancient civilisation might have been the *first* to recognise the number zero, there is general agreement that the Maya developed their symbol independently around 350 CE. Many other ancient civilisations had never even thought about the possibility of the number zero because they thought anything other than positive numbers was impractical or even vulgar. The Maya, however, saw the importance of using zero as a *placeholder*. They could see that while zero had no value on its own, it could be used to influence the value of surrounding digits based on their position. In any case, the sole use of the zero symbol in these early civilisations was to describe an absence of value in a particular place. There was no concept of zero being a number in its own right, with unique decimal and algebraic properties. While some scholars have used these arguments to discredit the Maya system, Robert Kaplan defends the Maya invention of zero saying, "The Maya counted as if their lives depended on it; and what they counted was time. Their starting-date for the universe would in our calendar be August 13, 3114 BC .... The Maya scrupulously recorded the dates of important events in terms of their zero day."[4] The importance of this day may have propelled for the Maya to create a unique symbol for a complex concept without outside influence.

Some early versions of Maya zero

[4] R Kaplan (1999) *The nothing that is: a natural history of zero*, Oxford University Press (p. 81).

Time trackers

The Maya had an obsession with numbers, astronomy and spirituality, which even regulated the calendars they used to keep track of time. Rather than tracking time using a single system, the Maya had different calendars for different purposes. The first of these calendars, called the Tzolk'in, was devised purely for carrying out sacred rituals. It was a 260-day cycle using 20 days and the numbers 1 through 13. It is not known why the Maya chose a cycle of 260 days. What is clear is that each day honoured a designated god to whom prayers and other offerings were made. For instance, the first day of every month was known as Imix and represented the god associated with a crocodile or water lily. It was a tradition in the Maya civilisation to name your child after the god associated with their day of birth on this calendar. Some descendants of the Maya in Guatemala carry on this tradition today.

Tzolk'in day names and glyphs

Imix   Ik   Akbal   Kan   Chicchan   Cimi   Manik   Lamat   Muluc   Oc

Chuen   Eb   Ben   Ix   Men   Cib   Caban   Etz'nab   Cauac   Ahau

While the Tzolk'in was important for sacred purposes, it was less useful for people like farmers who needed to keep track of changing seasons for their livelihood. For the common people, there was a second calendar called the Haab. The Haab was based on the solar cycle and was made up of 18 months with 20 days in each. There was an additional period of five days, called Uayeb, at the end of each yearly cycle. Anyone born during these five "unlucky" days was thought to be cursed.

Haab month names and glyphs

Pop Uo Zip Zotz Zec Xul Yaxkin Mol Ch'en Yax

Zac Ceh Mac Kankin Muan Pax Kayab Cumku Uayeb

Even though the Maya had calendars for different purposes, they combined elements of each into a third calendrical system known as the Calendar Round. It is not clear why they needed this third calendar, but it meant that each day effectively had a "double" name. It included a name and number from the 260-day cycle and a name and number from the 365-day cycle. Dates were always written in the same order: day number plus day name in the Tzolk'in and day number plus month name in the Haab. The set of interlocking wheels shown on the next page is one example of how the Maya cross-referenced the Haab and Tzolk'in when writing their dates. It shows the date as

with the number glyphs rotated from horizontal to vertical. This date reads "4 Ahau 8 Cumku". If the wheels are rotated one space each day, it takes 18,980 days (52 years) for the same date to reappear.

ACTIVITY 36:
The Maya Calendar Round

Learn more about the interplay of these different calendar systems by turning to page 179 and completing the activity.

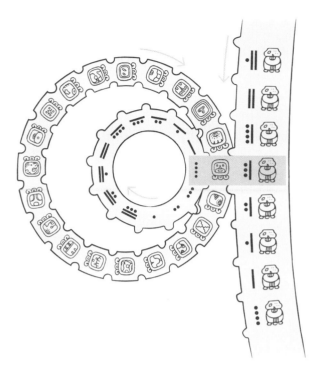

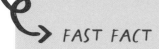

FAST FACT

The Maya Long Count Calendar was due to end at the completion of a Great Cycle of 13 *baktuns*, that is, 1,872,000 days after the perceived creation date. This corresponded to a date of 21 December 2012 on our modern-day calendar. For centuries, historians predicted that the universe would meet an apocalyptic fate on this date!

The Maya needed another calendar to keep track of longer periods of time. This calendar was known as the Long Count Calendar. It worked by counting the number of days that had passed since a date now identified to be the Gregorian (that is, our calendar) equivalent of 11 August 3114 BCE. For the Maya, this was the mythical creation date of the universe. The Maya people broke their long count dates into the units shown below. The units are all blocks of time that closely relate to the Maya number system used in everyday life. The Maya numerals, written vertically, would then indicate the total number of days.

Long Count – units of time

Maya name	Number of days
baktun	$20 \times 7{,}200 = 144{,}000$
katun	$20 \times 360 = 7{,}200$
tun	$18 \times 20 = 360$
uinal	$20 \times 1 = 20$
kin	$1 \times 1 = 1$

Tikal Stela 29

Tikal Temple 1, located in Tikal, Guatemala, is sometimes called the Temple of the Great Jaguar. It is a tomb for a ruler who lived approximately 1,400 years ago. The top of the temple is the ruler sitting on a jaguar throne. Jaguars are native to the region.

One important archaeological find is Tikal Stela 29 – a stone slab (or monument) that has the earliest surviving Long Count Calendar date from the Maya lowlands. The calendar stone was recovered from temple ruins at Tikal, an ancient Maya city located in the tropical rainforests of what is now Guatemala. Stela 29 was recovered in two distinct portions: the left part bears the sculpture of a king facing to the right, with the head of a

Tikal patron god clasped in his hand. On the right side, a series of Maya numbers and glyphs record the long count date 8 *baktun*, 12 *katun*, 14 *tun*, 8 *uinal*, 15 *kin*. The total number of days recorded on the stone is 1,243,615 (8 × 144,000 + 12 × 7,200 + 14 × 360 + 8 × 20 + 15 × 1). This represents the age of the stone in Maya long count terms. In Gregorian calendar units, the date of the stone is around 292 CE. Sounds about right!

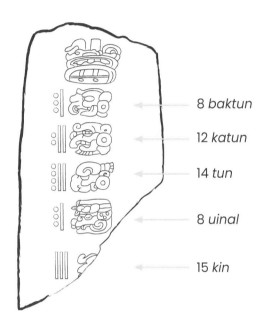

8 *baktun*

12 *katun*

14 *tun*

8 *uinal*

15 *kin*

ACTIVITY 37:
A time keeper for the long haul

Go to page 180 to complete the activity.

The Aztec civilisation

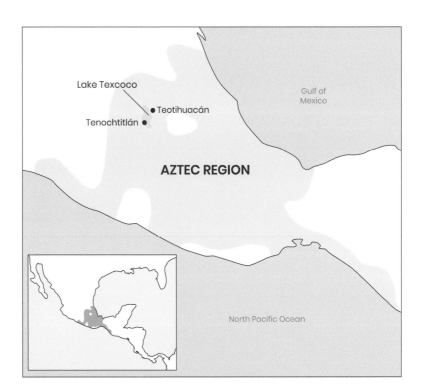

The Aztecs were the last of the Mesoamerican empires to wield power before the Spanish seized full control in 1519. The Aztecs had a brutal culture that relied heavily on wars to supply their priests with the captives they needed for their ritual sacrifices. There is evidence to suggest that up to 20,000 people were sacrificed each year during normal times.

Mexico Valley
c. 1519

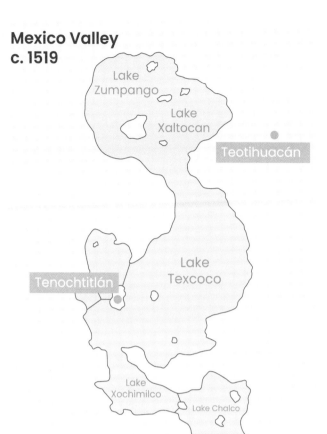

Lake
Zumpango

Lake
Xaltocan

Teotihuacán

Tenochtitlán

Lake
Texcoco

Lake
Xochimilco

Lake Chalco

The main region inhabited by the Aztecs was around these lakes. The main city was Tenochtitlán, on an island on the western side of Lake Texcoco. Today much of the region, including the water, is covered by Mexico City. A second major city was Teotihuacán where many original Aztec structures can be visited today.

The famous Pyramid of the Sun and the lesser structures were built about 2,000 years ago in Teotihuacán and pre-date Aztec rulers. The Pyramid of the Sun is the largest pyramid in the Americas. The Pyramid of the Sun and the nearby Pyramid of the Moon form the Avenue of the Dead.

Aztec pictograms

Much of what is known about the Aztecs comes from documents called "codices" (the singular is "codex"). These were written and kept by the Aztec high priests and tell of religion, magic, historical events and the origins and histories of dynasties. One of the most important codices is the *Codex Mendoza*. Although this codex was written for the Spanish court, it was based on accounts given by the Aztecs themselves. The accounts detail much of the history and culture of the Aztecs, including elaborate descriptions of their number system. Like the Maya, the Aztecs employed a vigesimal (base-20) number system. Their number system included four distinct symbols, each being a different power of 20. Unlike the Maya number system, the position of symbols was irrelevant in the Aztec system. Counting was simply a matter of adding the symbols together to find their overall value.

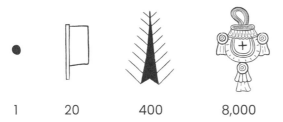

| 1 | 20 | 400 | 8,000 |

ACTIVITY 38:
Aztec pictographs
Go to page 181 to complete the activity.

While scholars differ in their interpretation of Aztec number pictures, it is generally thought that 1 represented a maize seed pod, 20 was a surveying flag, 400 was a feather, and 8,000 was a bag of incense. When numbers were used to show a quantity, they were always attached to a picture of the related object. For example, 20 shields and 8,000 deer hides were shown as pictured below.

20 shields   8,000 deer hides

With these symbols, the Aztecs recorded information relating to their lives, such as the number of prisoners captured in war and the trades and profits of merchants. At other times, these symbols prescribed quantities of goods that were demanded from the people of a conquered town. These "tribute demands" worked in much the same way as monetary taxes are paid to governments today.

Land surveying

The Aztecs were skilled land surveyors. The ability to measure the dimensions of land was crucial for determining the appropriate amount of tax or tribute to be levied. It also had a valuable economic incentive, since there was a need to maintain precise records of information relating to distribution, amount and quality of agricultural resources. In the Americas, the origins of land surveying are difficult to pinpoint, but two surviving records – the *Codex Vergara* and the *Códice de Santa María Asunción* – prove that a sophisticated system was used by the Aztecs prior to European contact.

Aztec surveyors recorded side lengths of hundreds of agricultural fields using a standard linear measure known as a "land rod" or *tlalcuahuitl* (T). A standard land rod is equal to about 2.5 m in the modern metric system. Whenever a measurement did not match a precise number of land rods, the Aztecs added symbols to indicate the remaining length (which was less than one rod). The "additional" symbols were represented by an arrow and different parts of the human body. The table below shows the various units of measurement that were used by the Aztecs for land surveying.

Glyph	English	Fraction of a land rod (T)	Metric equivalent (m)
	Arrow	$\frac{1}{2}$	1.25
	Arm	$\frac{1}{3}$	0.83
	Bone	$\frac{1}{5}$	0.5
	Heart	$\frac{2}{5}$	1.0
	Hand	$\frac{3}{5}$	1.5

The Aztecs would not, of course, have understood these additional units as representing fractions of a land rod. While each of the sub-units may have served essentially the same *role* as a fraction, to the Aztecs they were considered stand-alone entities similar to the way we think of inches, seconds or minutes.

Historians have a range of theories to explain the choice of symbol for each unit of measurement. It is believed that an arrow represents the length from the shoulder to the hand, like an archer holding a taut bow. The heart is thought to be a measure of the length from that organ to the tip of the hand, and a hand might represent the distance from one outstretched hand to the other – just as an English foot is the measure of an "average" man's foot size.

These units had a far broader application than simply recording the linear dimensions or perimeter of agricultural fields. An analysis of the *Codex Vergara* shows that the Aztecs also came up with several algorithms to calculate land area. These algorithms showed an understanding of multiplication, division and basic principles of geometry. By analysing the codex, one research project led by the University of Mexico (in collaboration with the University of Wisconsin, USA) discovered schematic outlines of 367 individual fields. More than half of these fields had areas that matched the basic mathematical rule of length multiplied by width ($l \times w$). However, not every field in the codex had the shape of a perfect right-angled quadrilateral. Sometimes the shape of a field was more complex. When dealing with irregular four-sided lots, the Aztecs had to use a different approach. This often involved multiplying the average of two opposite sides by an adjacent side. The Aztec measurement system was complex for its time period, and it allowed the surveyors to come up with very accurate and consistent estimations of land area.

One interesting aspect of the Aztec approach was their use of square *tlalcuahuitl* units to record land area. Area was not measured by some physical method but was computed by counting the number of individual square units enclosed within a space. When the Spaniards conquered the Aztec empire at the beginning of the 16th century, they were unable to interpret the Aztec measurements. Instead, they adopted an indirect method that quantified the amount of maize or wheat produced from the land that was farmed.

The following images are based on those found in the *Codex Vergara*. They are highly mathematical and relate to the dimensions of plots of land. The Aztecs used one system to find the perimeter of the fields. Another system was used to record the areas of those fields. The image below represents the holdings of a 16th century Mexican landowner.

Image A shows a set of fields with their side lengths included. The small symbols along the inside edge of the rectangular fields record the lengths of the individual sides in Aztec measurement units (*tlalcuahuitl*). A solid dot represented 20 T, a vertical line represented 1 T, and a set of five lines grouped together represented 5 T. Shorter-than-standard distances include a hand (here appearing in both fields) and an arrow (second field). For example, the dimensions of the right-hand field are, clockwise from the top, 35, 34 + hand, 29 + arrow and 39. The perimeter was the total distance around the field. Adding the individual side lengths together, the perimeter of the right-hand field would be read as slightly more than 138 T.

ACTIVITY 39:
Aztec area activity
Go to page 182 to complete the activity.

Image A

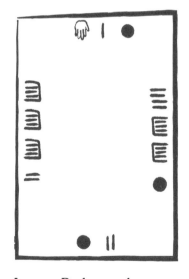
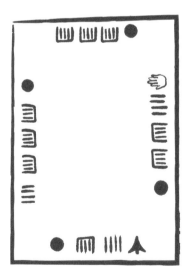

Image B shows the same fields with their corresponding areas. The areas are written in Aztec place notation with the twenties place in the centre and the ones place in the tab on the upper right. To read the area (given in square *tlalcuahuitl*), the symbols in the centre of the rectangle were multiplied by 20 and then added to the units in the upper right-hand tab. For example, the area of the second field would be read as $59 \times 20 + 12 = 1{,}192$ square units.

Image B

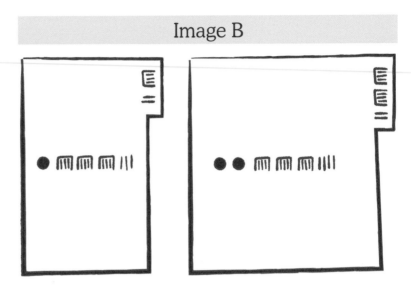

In certain instances, fields did not always have a perfect rectangular shape. Some of the images in the codex show land plots shaped more like trapeziums. In these cases, the Aztecs had to slightly modify their method in order to read the land area. It can be inferred that they read the area as the product of the averages of the opposite sides. For example, the field shown here has sides that measure, clockwise from the top, 6, 49, 23 and 52 *tlalcuahuitl* units. The area of the field was taken by finding the average of the opposite sides, and multiplying these averages by each other. In square units, the area of the plot was therefore $(6 + 23)/2 \times (49 + 52)/2$, which is approximately 732 square units.

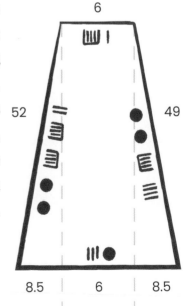

Aztec sun watchers

Many ancient cultures designed their buildings and monuments according to their sense of spirituality and advanced understandings of astronomy. One example of this purposeful use of architecture is the Templo Mayor, one of the largest temple-pyramids built in the Aztec capital of Tenochtitlán.

The Templo Mayor was made up of four distinct terraces and two shrines built on the temple top. These shrines were dedicated to the gods Tlaloc and Huitzilopochtli, the gods of rain and war, respectively. It is believed the building was erected in the spot where Huitzilopochtli gave the signal that the Mexican people had found their promised land. As well as a place of worship,

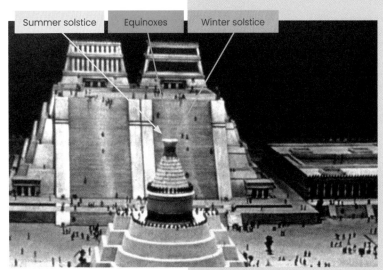

This model of the observatory and pyramid shows the position of the rising sun at the key astronomical times of the year.

Reconstruction of Tenochtitlán, the central city of the Aztec civilisation, with a view to the sacred precinct. The pyramid was the central focus of the city with the surrounding palaces and lodgings for priests and other high-ranking officials. There are very few remains of the city as most of Tenochtitlán is now buried under Mexico City's main square.

the Templo Mayor was also a site of regular and brutal human sacrifices. The blood of victims was thought to feed and appease the two great gods to whom the temple was dedicated.

Despite its history of blood-shed, the Templo Mayor also served as an astronomical observatory. The design of the building suggests the Aztecs understood that the time of year and the activity of the sun were related. The temple is oriented so that the two staircases leading up to the shrines face west. The Aztecs understood this to be the place where the sun descended the evening sky into the underworld. They also had a special way of determining the timing of both yearly equinoxes. The way the building was positioned meant that the rising sun on each equinox would shine directly between the two shrines on top of the pyramid.

Aztec sun stone

Like the Maya, the Aztec people of Mexico also had a year of 365 days. Their year was divided into 18 months of 20 days. The five extra days were called "empty days".

The Aztecs recorded each day of the month on the sun stone, known in Spanish as Piedra del Sol. As shown in the drawing of the sun stone below, there is a different symbol for each day.

This circle shows the 20 symbols for the days.

The sun stone is one of the most enduring symbols of Aztec iconography. It is believed the stone was carved sometime in the late 15th century. Naturally, the stone was dedicated to Huitzilopochtli, who was the sun god as well as the god of war. It was a massive carving, measuring nearly 1 m thick, more than 3.5 m across, and weighing almost 25 tonnes. The stone was carved from basalt, a solidified lava, this being an area where volcanoes were common. But then it was lost – buried under the central square of Mexico City – for over 300 years.

In 1790, renovations began on the central square of Mexico City. The Aztec sun stone was unearthed, renewing interest in the long-lost Aztec civilisation that once dwelled there. For a while the stone was on display in the Metropolitan Cathedral, but in 1855 it was moved to the National Museum of Anthropology in Mexico City, where it remains to this day.

The famous sun stone is a brilliant combination of artistry and geometry. It reflects the Aztec understanding of time and space as wheels within wheels. The detailed surface of the stone combines the understanding of the gods the people had created over the centuries as well as their observations of the heavens.

Gambling games

One of the most popular games developed by the Aztecs was called patolli. It involved tossing a form of dice made from cacao beans and moving pebble counters around an X-shaped board. While the game is still played recreationally today, the main reason the Aztecs played it was to gamble. Mostly, the participants would bet everyday items such as blankets and food, but the stakes were sometimes much higher. When played by the Aztec nobility, players would often put very expensive possessions, like their homes, on the line. The game could even be played with participants betting their civil liberties and status in society. In extreme cases, the Aztecs used the game as an excuse for assassination.

Patolli was a game of both chance and strategy. The only certainty was that both players had to enter the game betting the same number of items. Once the agreed number of betting items was settled, the players were at the mercy of the god of games, Macuilxochitl. There was no turning back!

The game of patolli was very popular among the Aztec people. In this game, two players are being observed by two neutral spectators with the god of games (Macuilxochitl) looking on. Beans were used for counters and moved around the board. In addition, beans were used as dice with a hole drilled on one side of each bean. When six beans were tossed, the number of holes face up indicated the number of spaces to be moved. Because there are two possible outcomes, this type of game is an early example of a binary situation. When the Spanish arrived, the game was banned and playing was punishable with death.

The typical number of items to bet was six, since the players started the game with this number of markers. Each time a marker successfully completed a circuit around the board, the opponent was required to hand over one of their items. A round was completed once a player got all six of their markers from the starting queue to the end of the board, through a series of specially marked squares. However, the game itself was not won until a player had taken hold of all of the opponent's items. Sometimes, this meant multiple rounds had to be played. The players continued to battle it out until there was a definitive winner and loser.

Aztec artists

ACTIVITY 40:
The artsy Aztecs

Go to page 183 to
complete the activity.

ACTIVITY 41:
**Maya and Aztecs
word match**

Go to page 184 to
complete the activity.

Aside from being a very practical civilisation, the Aztecs also had a distinguished sense of creativity. Art was a central part of Aztec life. Its influence permeated many areas of culture, from sacred offerings through to jewellery worn by the Aztec nobility. Another example was the pattern they used to decorate their shields for use in warfare. The picture below shows a typical shield design. It shows a special kind of mathematical symmetry known as "anti-symmetry". The gold and brown areas are symmetrical shapes, but the colours are opposite on the colour wheel.

It is not known why the Aztecs chose such a perfectly balanced design for their shields. Maybe it was to shift the enemy's focus away from the attacker, or perhaps it was yet another instance of the Aztecs tapping into their inventive ways.

On this shield, the gold and brown regions have the same outline. If the gold region is rotated it will therefore fit exactly on top of the brown outline. If both regions were gold, the design would be described as rotationally symmetrical. However, in this situation, with the regions being in "opposite" colours, it is described as anti-symmetrical.

Summary

Mesoamerica was the historical region comprising most of the modern-day countries of Central America. The main civilisations from this area include the Maya and the Aztecs. They had many shared cultural, topographical and architectural characteristics.

Numbers

- The Maya came up with one of the earliest symbols for the number zero.
- The Maya designed two number systems: one for common people, one for priests.
- Mesoamericans used a base-20 positional number system.

Geometry

- The Aztecs devised a unit of measure to help them measure the linear dimensions and surface areas of agricultural fields.
- The Aztecs designed their shields with anti-symmetry patterns.

Measurement

- Mesoamericans devised different calendars for measuring time.
- The Maya devised a calendar to keep track of time over significantly long periods (known as the Long Count Calendar).
- Mesoamericans had units of measurement for land surveying.

Probability

- The Aztecs designed a gambling game that was based on a mix of strategy and pure chance.

The Mesoamerican legacy

Mesoamericans saw how time could be measured in both the short and long term, and they changed the face of arithmetic by introducing the concept of zero value. The Aztecs introduced the concept of measuring area by square units.

Activity 31: *Mapping the Mesoamericans*

Study the map below.

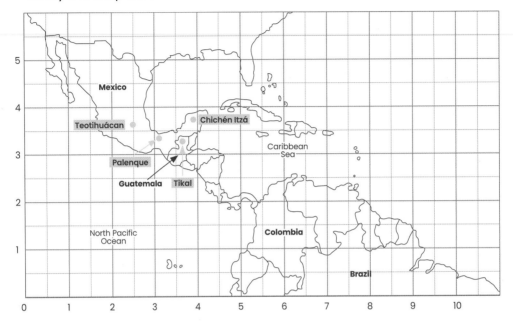

1. Write the name of the historical landmarks located closest to each of the following coordinates. Write the name of the modern country where these landmarks are located.

Landmark	Name	Modern country
a. (4, 4)		
b. (2.5, 3.5)		
c. (3.5, 3.5)		
d. (3, 3.5)		

2. Write the names of the countries you know that are shown on the map. Use the space below to write the country names.

_____

_____

_____

Activity 32: *Magnificent Maya*

1. Answer these questions about the year.

    a. How many days in one year (not a leap year)? _____

    b. Write the month that each of these events occurs in the Northern Hemisphere (where El Castillo is located).

i. Spring equinox _____   **ii.** Summer solstice _____

iii. Autumn equinox _____   **iv.** Winter solstice _____

    c. Write an estimate of the number of days between each of the events.

_____

2. What geometric shape best describes El Castillo?

_____

3. El Castillo has four symmetrical staircases, each with the same number of steps. There is one additional step at the summit. In total, how many steps does El Castillo have? _____

4. Why do you think the Maya would chose to build their staircases with four sides of 91 steps? _____

_____

Activity 33: *Cracking the Dresden Codex*

1. Work out the values for each stone marking below. Write each value.

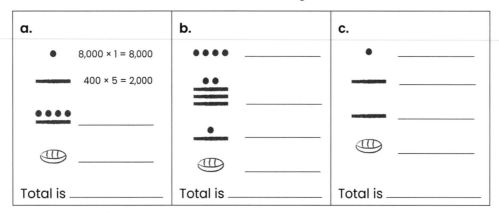

a.

●   8,000 × 1 = 8,000

━━━   400 × 5 = 2,000

●●●● _____

⬭ _____

Total is _____

b.

●●●● _____

●●
━━━
━━━ _____

●
━━━ _____

⬭ _____

Total is _____

c.

● _____

━━━ _____

━━━ _____

⬭ _____

Total is _____

2. Three different Hindu–Arabic numbers are given below. Work out the value for each place and then write each number in Maya form.

Groups of:	555	4,573	34,866
8,000s			
400s			
20s			
1s			
The Maya number symbols			

Activity 34: *One number, two meanings*

The Maya high priests invented a separate number system that they used to record astronomical cycles and other units of time. It was similar to the ordinary number system, except that the third place was counted in units of 360 (18 × 20) instead of 400 (20 × 20).

COMMON PEOPLE		HIGH PRIESTS
5 × 400	▬	5 × 360
7 × 20	•• ▬	7 × 20
18 × 1	••• ▬▬▬	18 × 1
= 2,158		= 1,958

1. Why do you think the Maya high priests preferred a grouping of 360 units (rather than 400) in the third place? _____

2. How would a Maya priest have written these numbers of days?

a. 180 days	**b.** 400 days	**c.** 500 days

3. Would the common people have written any of the above numbers differently? If yes, show how:

a. 180 days	**b.** 400 days	**c.** 500 days

Activity 35: *A place for zeros*

The Maya seem to be the only early civilisation that had a zero early in their history. They saw great utility in using their zero symbol as a *placeholder*. Without a zero, numbers could take on vastly different meanings.

1. The chart below contains a list of three-digit numbers. Follow the instructions to fill in each column.[5]

Steps:
1. In column A, write the number fully in words.
2. In column B, rewrite the number in symbols but remove the zero.
3. In column C, write the new number in words.
4. In column D, record how necessary the zero is to preserve the original value (N – Necessary, H – Helpful but not necessary, U – Unnecessary).

	A	B	C	D
290	two hundred and ninety	29	twenty-nine	N
029				
0.029				
0.92				

2. Use the digits 5, 7, 0, 0 to write *two* numbers in which:

a. both zeros are necessary _____ _____

b. one zero is necessary, while the other is merely helpful

_____ _____

c. both zeros are unnecessary _____ _____

Activity 36: *The Maya Calendar Round*

1. Use the picture to help answer questions about the Maya Calendar Round.

   a. How many numbers are on the smallest number wheel? _____

   b. How many images are shown on the middle wheel of day names?
   _____

   c. When will the two small wheels be back to the starting point? _____
   (Hint: the answer is the product of the number of images on each wheel.)

   d. The large wheel has 18 × 20 + 5 images. How many images are on that
   wheel? _____

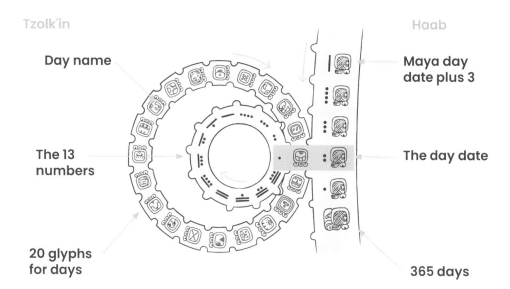

Tzolk'in

Haab

Day name

Maya day
date plus 3

The 13
numbers

The day date

20 glyphs
for days

365 days

2. The day shown is 1 Kan 2 Pop. Look how the wheels turn. Write the numbers for

   a. the next date: _____ Pop

   b. the date in two days: _____ Pop

   c. the date in three days: _____ Pop

   d. the date for the day before: _____ Pop

3. A full Calendar Round is approximately 18,980 days. What is the
approximate number of years (365 days) in this same time? _____

Activity 37: *A time keeper for the long haul*

The Maya broke their long count dates into the following units of time.

Long Count – units of time

Maya name	Number of days
baktun	20 × 7,200 = 144,000
katun	20 × 360 = 7,200
tun	18 × 20 = 360
uinal	20 × 1 = 20
kin	1 × 1 = 1

1. Using the table above, estimate the number of years equivalent to:

    a. 2 *katun* _____

    b. 13 *baktuns* _____

    c. 15 *tuns*, 4 *uinals*, 10 *kins* _____

2. The diagram below represents numbers on a Maya stone. How many days does it add up to?

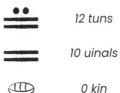

    3 *baktuns*

    11 *katuns*

    12 *tuns*

    10 *uinals*

    0 *kin*

Total = _____ days

3. The diagram below shows a gigantic number of days. Work out and write the total shown.

_____ *baktuns*

_____ *katuns*

_____ *tuns*

_____ *uinals*

_____ *kin*

Total = _____ days

Activity 38: *Aztec pictographs*

1. Work out and write the amount of each
of the captured goods shown below.

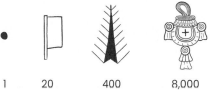

1	20	400	8,000

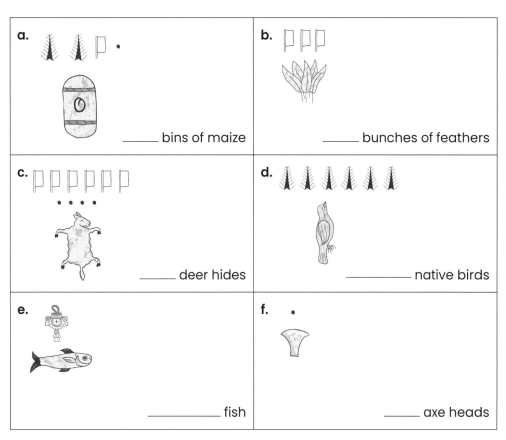

a. _____ bins of maize

b. _____ bunches of feathers

c. _____ deer hides

d. _____ native birds

e. _____ fish

f. _____ axe heads

2. Use Aztec images to show each of these numbers.

a. 320	b. 3,200	c. 32,000

3. What pattern do you notice? _____

Activity 39: *Aztec area activity*

The unit of length used by the Aztecs was called a *tlalcuahuitl* (T).
Their measurement unit for area was a square *tlalcuahuitl* (T²).

When they wrote lengths or area, they used a ● for 20 units
and | for one unit. The symbol 𝖕 meant five units.

●	= 20 T
|	= 5 T
𝖕	= 1 T

1. Write the number represented by these Aztec measurement symbols.

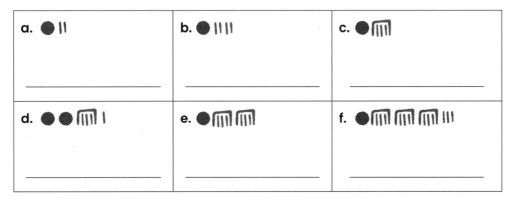

a. ● ||	**b.** ● ||||	**c.** ● 𝖕
_____	_____	_____
d. ● ● 𝖕 |	**e.** ● 𝖕 𝖕	**f.** ● 𝖕 𝖕 𝖕 |||
_____	_____	_____

2. When the Aztecs wrote the area of land, the number drawn in the middle of the field was multiplied by 20. The number on the tab was added to the first amount. Complete each number sentence to find the total area of each piece of land.

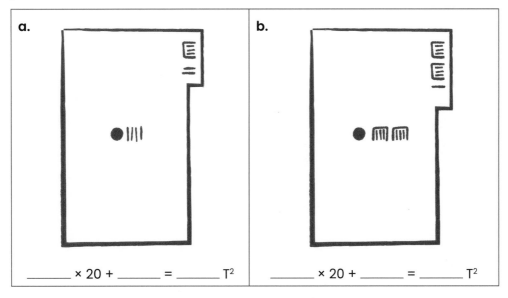

a.

_____ × 20 + _____ = _____ T²

b.

_____ × 20 + _____ = _____ T²

Activity 40: *The artsy Aztecs*

1. Use two different colours on each pair of shapes below to create designs that display the following.

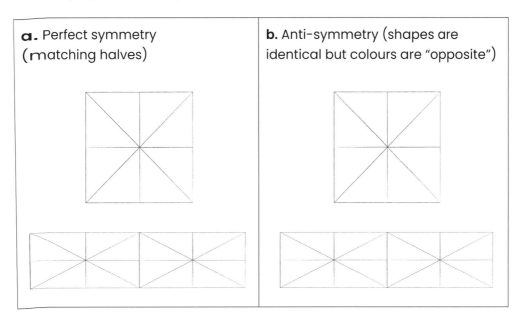

a. Perfect symmetry (matching halves)

b. Anti-symmetry (shapes are identical but colours are "opposite")

2. These are two Aztec tiling outlines. Use two colours to shade each design to show anti-symmetry.

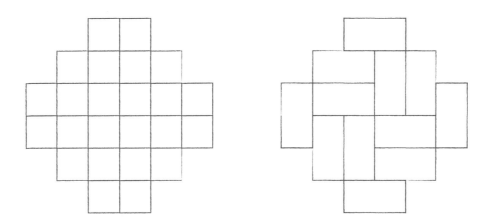

Activity 41: *Maya and Aztec word match*

1. Draw a line from the description to the single word that matches.

The Long Count unit equivalent to 144,000 days.	The Tzolk'in day glyph that comes immediately after Ahau.	Country in Asia with the oldest recorded reference to zero-dot symbol.
A piece of defensive equipment used by Aztecs during combat.	The function of zero in the Maya number system.	The sacred Maya calendar composed of 260-day cycles.

Vigesimal	*Baktun*	**Tzolk'in**
El Castillo	**Cambodia**	**Shell**
Imix	**Anti-symmetry**	**Placeholder**
K'ank'in	**Haab**	**Shield**

A positional number system with a base of 20.	The object from the environment that represents the Maya zero.	A pattern where the shapes are identical but the colours are "opposite".
A Mesoamerican step pyramid situated on the Yucatán Peninsula.	The month that follows Mac on the Maya Haab calendar.	The second Maya calendar that common people used and was based on the solar cycle.

Working space

CHAPTER FIVE

The Inca

When? Where? How?
Who? Why?
I wonder, wonder,
wonder, wonder.

Where can we find ancient buildings that have remained standing even after the strongest earthquakes?

Have you walked on a catenary? Where did you do that? What do you remember?

How could you work with numbers, keep good financial records and not write any symbols?

Have you ever been to Machu Picchu? If so, what country were you visiting?

How did indigenous people of South America influence Europeans to stop using Roman numerals?

The Inca: *calculating numbers without symbols*

The Inca civilisation was located along the west coast of South America, bordering the Andes mountains. There are many myths, legends and stories that describe the ancient Inca. Based on information provided by the 16th century Spanish explorer and mathematician Pedro Sarmiento de Gamboa, it seems the Inca belief system was based on the cosmos and study of the stars in

Most people are aware of the famous World Heritage–listed landmark in South America even though they may not know its name, the history behind it or where it is located. Machu Picchu is a marvel, and the construction techniques make it a mathematical marvel. The site also contains features to study astronomy that incorporate mathematical ideas. This picture is taken standing on Machu Picchu, which means "old mountain". The tall mountain in the middle of the photograph is Huayna Picchu, which translates as "young mountain".

the Milky Way. According to Sarmiento, this is one of the reasons for building special sites, such as Machu Picchu, that were close to the "heavens".

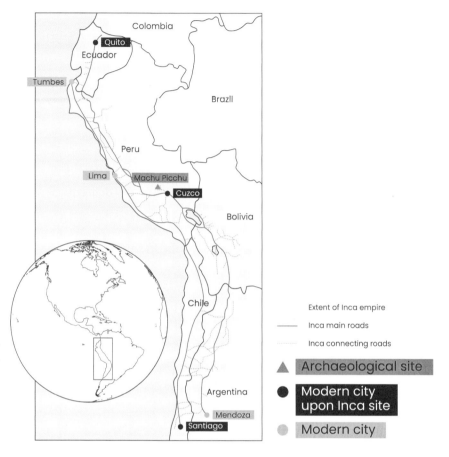

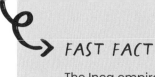
The Inca empire stretched from just north of Quito, Ecuador, to near Santiago, Chile, in the south. This was a distance of over 5,000 km. Cuzco was positioned at about the halfway point between the two extents of the civilisations. The longest north–south distances in the USA and in Australia are far shorter, as shown in these charts. Teams of Inca runners carrying quipus could run up to 320 km/day. Each runner would run at top speed for up to 2 km.

Quito, Ecuador	Portland, Maine, USA	Cape York, Queensland, Australia
↓ 5,000 km	↓ 2,600 km	↓ 3,700 km
Santiago, Chile	Miami, Florida, USA	Melbourne, Victoria, Australia

Pre-Inca records can be dated to 6,000 years ago when hunter gatherers settled in the Cuzco region of modern-day Peru. Officially, Cuzco became the centre of the Inca civilisation about 1,000 years ago. Over the next 450 years there was a rapid expansion of the empire that eventually reached modern-day Ecuador in the north and Chile and Argentina in the south. Also, during this time, the ruler Pachacuti Inca Yupanqui (1438–1471) built Machu Picchu high in the Andes mountains just north of Cuzco. From 1471 to 1493, under the reign of Topa Inca Yupanqui, the Inca empire doubled in size, and it reached its greatest extent in 1530. It is believed that, at the time, approximately

10 million Inca lived in a region that spread over 5,000 km from north to south. However, we do not have any written evidence to support this because, unlike other early civilisations, the Inca did not develop any form of a written language. What we know today about the Inca people and their culture comes from the Spanish soldiers, priests and administrators who invaded the empire in 1532 and, in so doing, destroyed it.

There is uniform agreement that the Inca were very highly organised. Their ability to establish an organised society was assisted by two features: a system of roads that enabled rapid travel, and communication facilitated by the "quipu". The quipu was a seemingly sophisticated yet simple record-keeping device. An alternative to written records, quipus were systems of knots tied on different-coloured cords that were used to keep detailed accounts of all numerical aspects of the population and general economy. The road system and quipus allowed information to move quickly throughout the vast civilisation, enabling Inca officials to maintain control of the empire.

It is interesting to note that all other early civilisations had some form of written language, based on symbols that were *pictographic* (pictures represent ideas – Babylon, Egypt, Mesoamerica), *logographic* (characters represent words – India, China) or *alphabetic* (Greek). Given they had no written language, the achievements of the Inca are truly amazing.

ACTIVITY 42:
Keeping in contact?

Go to page 210 and complete the activity to explore the area where the Inca lived.

A visual number system – the quipu

The quipu aided in the establishment of an Inca empire that covered a vast distance. Control, consistency and conformity were achieved through information that enabled decision-making. The quipu facilitated quick collection, storage and transport of data to update government records, which could be analysed by Inca officials.

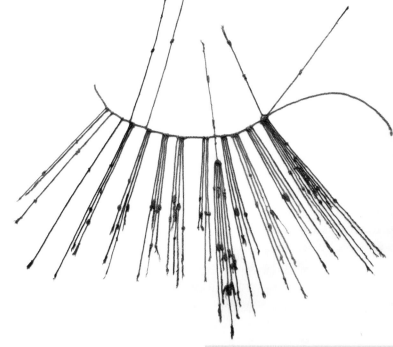

The Inca would have made thousands of quipus, but most were destroyed by the Spanish when they conquered the Inca civilisation. The Khipu Database project at Harvard University has reported that today 751 exist across the globe.

The Amazing Beginnings of Mathematics

191

When the Inca conquered a region, their accountants would go in and count up all the different resources: the people of different genders and ages, the animals, the streams, the fields, the mining possibilities, the fishing areas and anything else that was of value. Different-coloured cords were used to indicate the type of information that was recorded. This information was sent back to Cuzco where decisions would then be made about managing the area. The information was predominantly numerical: knots of different lengths were tied on pendant cords hanging vertically from a main cord across the top of the quipu, as shown on page 191. In some cases, secondary cords were attached to the pendants.

The individual knots were arranged vertically in groupings that suggested a base-10 number system. This was like a horizontal version of the abacus used in other parts of the world, as shown below. One of the reasons that the Inca developed the quipu rather than a written system was the geography of their civilisation. Information needed to be communicated to Cuzco, the centre of the Inca empire, from vast distances. Written data would have been nearly impossible to transport given the lack of material such as paper or papyrus.

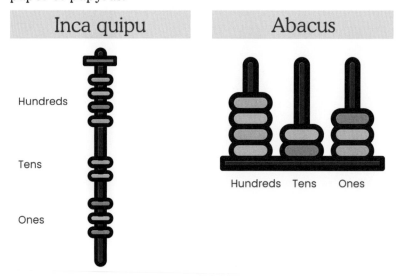

Knots on pendants or cords were used to show all numerical values. On a quipu, the greatest place value for a number was positioned at the top of the cord. Three types of knots indicated the value in each place: the figure eight knot, the long knot and the single knot. The special knots for the ones place helped to establish this as a beginning reference point for a number.

A **figure eight knot** was used only in the ones space and represented one in that place.

A **long knot** was used only in the ones place to indicate two to nine in that place. It was a continuous looping of turns around the pendant cord, so it was necessary to count the number of loops.

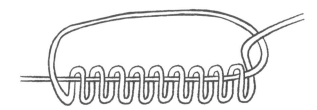

Single knots were used only in the tens, hundreds, thousands and greater places. These knots were separate, so it was necessary to count the number of knots to know the value in those places.

The Inca did not have a knot for zero. In the ones space, the two types of knots indicated a value, so in the absence of either type of knot, the Inca knew the value in that place was zero. For the other places (tens and greater), the best they could do was leave the "space" empty. This could, of course, be confusing and lead to errors. Using this information, can you tell what numbers are represented on this quipu?

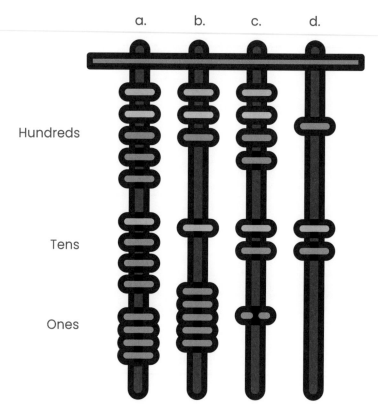

See the answers on the next page.[6]

Quipus could be challenging to read, especially when they were recording all sorts of different things, such as you would find in a newly conquered town. We have seen that placement (from top to bottom) gave the value of the number. The colour of the quipu identified what was being counted (such as population, grain or animals). The thickness of the cord indicated the importance of the number. For example, a cord for the total contents of grain storage warehouses in the region would be thicker than one for grain in a single village. Finally, the arrangement – with

~ᛗ
ACTIVITIES 43 & 44:
Reading and representing numbers

Complete the activities on pages 212 and 213 that give you practice interpreting and then showing numbers on an Inca quipu.

a cord attached to the side of a primary cord – provided data for connected pieces of information.

Quipu specialists called *quipucamayocs* were trained as keepers of this special device. These specialists were like ancient accountants who both created and deciphered the quipus. *Quipucamayocs* were located in provinces where they would collect all of the information and send it back to Cuzco to be tabulated and recorded on other "master" quipus.

Just before the Spanish arrived, there was a civil war within the Inca empire. Two rival brothers, Waskar and Atahualpa, fought for control, with the latter winning the battle for leadership. During the skirmishes, information was vital, so many quipus were destroyed. When the Spanish arrived in South America, they were uncertain about the purpose of the quipu. They became suspicious because the Inca referred to the knots on the quipu like they might read some text. As a result, the suspicious Spanish destroyed more (thousands) of the remaining quipus. The destruction during these two events resulted in fewer than 800 quipus surviving today, many of which are in museums outside South America.

[6] Answers for the numbers on the quipus above: a. 544, b. 315, c. 421, d. 120.

Calculating with a *yupana*

CÖTADOR MAIORÍ TE30RERO
TAVANTIN:SVIO:QVÍPOC
CVRACA·CON DOR·CHAVA

con taxdor ytqoune   con ta ber

Felipe Guáman Poma de Ayala (c. 1535 to c. 1615) was an indigenous Peruvian author and illustrator who prepared a 1,189-page written history with 398 drawings of the Andean civilisation. He intended this for King Phillip III of Spain. A famous image from the book is shown here with two mathematical tools used by the Inca.

In addition to recording numbers on the quipu, the Inca employed a *yupana* to calculate "sums" that appeared on the quipu. The Spanish found many *yupana* in the Andes, but the method for using the *yupana* was unclear. Some of the challenges faced in solving the mystery of the *yupana* relate to its variable design. Generally, the *yupana* was made of stone with

A drawing from a series completed by Felipe Guáman Poma de Ayala, a Quechua (one of Peru's indigenous peoples) who lived during the time of the conquest by the Spanish. The illustrations show some of the important facets of Inca life. The person in this picture is holding a quipu and standing beside an image of a *yupana*, a calculating tool unique to the Inca. This illustration is housed in the Royal Library, Copenhagen, Denmark.

rows that aligned to the places on a quipu. So starting from the bottom, the rows represented the ones, tens, hundreds and finally thousands places on a quipu. However, the few examples that were found varied greatly, which made it difficult to decode the steps used to perform mathematical operations. There is also the possibility that the steps for adding (the main mathematical operation the *yupana* was used for) varied from one part of the empire to another.

Guáman Poma lived among the Inca for some time and his account of how the *yupana* was used is the most reliable. He and his Spanish compatriots worked out how to add on a *yupana*. There is no evidence that the Inca had developed methods to carry out other mathematical operations.

Within the past few decades, additional analysis of Guáman Poma's works and archaeological research has provided further information about the *yupana* and how it was used. The analysis has shown that pebbles were placed in the first three columns of the *yupana* to show one of the numbers to be added. Pebbles placed at the right of this abacus-like device showed the other number. The images on the next page summarise the steps to add 315 and 439.

The number 439 is shown on the *yupana*. The pebbles to the right of the yupana show 315. The pebbles off the *yupana* (315) are moved one at a time onto the *yupana*, starting in the ones row.

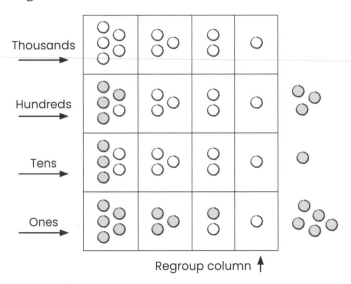

When 10 pebbles cover the spaces in the first three columns of the ones row, these pebbles are exchanged for one pebble in the regroup column of that row. It is then moved to the next row above (the tens).

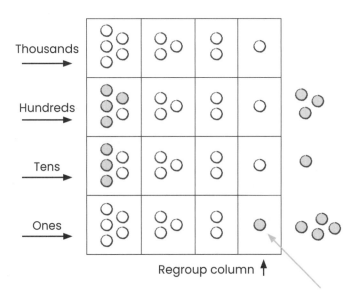

The remaining ones are then moved onto the *yupana*.

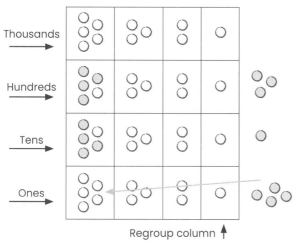

Regroup column ↑

The processs is then repeated in the tens, hundreds and thousands rows as necessary. Further analysis of the 500-year-old manuscripts might reveal more information about the exact steps used with this unique calculating tool.

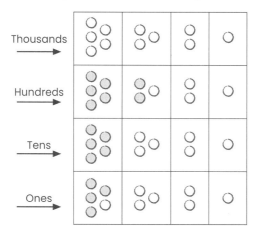

The quipu (decimal record system), *yupana* (decimal calculating tool) and the numerical literacy of the Inca helped the Spanish profit commercially. It is theorised that the sound commercial methods learned from the Inca helped the Spanish progress ahead of other countries in Europe. The Spanish probably gained this advantage while the rest of Europe was still struggling to rid themselves of the cumbersome Roman numerals. It is even possible that the *yupana* helped to accelerate the use of the base-10 Hindu–Arabic number system that had been introduced 300 years earlier but had not yet been accepted in some countries in Europe.

ACTIVITIES 45 & 46:
Adding with a yupana

Go to page 214 and complete the steps to add numbers on a *yupana*. Use counters with the *yupana* picture on page 215 to help.

Inca – the master builders

The Inca were builders. They constructed their roads and buildings to remain intact even during earthquakes. This finely finished wall is the outside of a temple building in the Inca capital of Cuzco. For temples and other special buildings, the stones were nearly identical and formed neat tessellating patterns. The large stones were notched so they would interlock and therefore be difficult to slide out of place. The walls usually had a slight taper of about five degrees inward, which also helped to strengthen the building.

The Inca were master builders. These photographs were taken in Cuzco, the most important and central city in the Inca civilisation. The walls in this picture show the fine detail the Inca could achieve where it was important to have a good-looking finish.

When the Spanish arrived in Cuzco, they added to some of the Inca-constructed buildings. When strong earthquakes occurred, many of the sections built by the Europeans were destroyed while the Inca portions of the structures remained intact. This was due to internal interlocking features of the blocks that could not be seen, as shown here. The tight tessellations, inward sloping walls and unique interlocking design combined to give exceptional strength to the Inca buildings.

Stones were cut to interlock and withstand the earthquakes that are a part of the region. There is a total absence of mortar to fill gaps that might be seen between the massive stones. There weren't any gaps to fill!

Buildings that seemed to have less importance did not show the same fine detail of stonework. Homes, workshops and schools were erected with blocks that could be described as "almost rectangular". The main characteristics of these buildings were the trapezium features such as the overall shape of the structure as well as the windows and doorways. To be more specific, the openings were isosceles trapeziums – the sides are equal lengths.

The advantage of using a trapezium for the openings was the strength provided. The wider base of the trapezium provided structural advantages over a rectangle used for a door or window today. Even small niches in the walls, as shown through the doorway below, were made with the stronger shape.

The trapezium was a feature of the openings and very often the overall shape of buildings such as those seen in Machu Picchu. It is a naturally strong shape that does not require extra support.

ACTIVITY 47: Analysing angles

Go to page 216 and complete the questions about angles found in 2D shapes.

The third and lowest standard for building was used for outside retaining walls. These were typically required to terrace the hillsides. The stones were usually irregular but carefully carved so there were no gaps, as shown on the opposite page. The large stone shown on the next page has 12 sides of varying lengths and 12 unequal angles (it is an irregular dodecahedron). The Inca didn't worry about the measure of the angles. They simply carved the stones to fit what was already in place, without the need for mortar. This resulted in a mosaic effect rather than a pure mathematical tessellation. Either way, there was a lot of careful crafting involved.

This stone is used in many photographs to show how the Inca could arrange odd-looking shapes in a wall with minimal carving and no mortar even though the angles were awkward. The outline of the stone provides a good opportunity to review names for different angles.

The Inca built approximately 40,000 km of roads throughout their empire. The road work was commenced by the Wari, Tiwanaku and Chimú peoples before the establishment of the Inca empire, so some sections are more than 1,000 years old. The road stones were smooth and laid in a mosaic manner with the gaps filled with a soil mortar. They were said to be better built than the roads constructed by the Romans.

FAST FACT

Some definitions for a trapezium state that "a trapezium is a quadrilateral with at least one pair of parallel sides". Mathematicians prefer this definition because rules for finding the area of quadrilaterals can be related across a greater range of shapes.

Two main roads extended north to south with Cuzco at the centre. One road followed the coast, while the second stretched through the Andes. Some sections of the latter roadway were more than 4,500 m above sea level with many twists and turns to create gradual slopes up and down the mountains. The Inca did not have wheels; they used llamas and porters to carry goods from one place to another. The construction enabled the Inca to move their armies relatively quickly, but, more importantly, the system allowed individual runners, called *chasquis*, to carry messages rapidly. Post houses were strategically spaced along the route, which enabled the *chasquis* to race at top speed and traverse long distances in a relatively short period of time. They carried a quipu, which contained valuable data and was easy to hand from one runner to the next as they raced along their stone roadways.

ACTIVITY 48:
Walking the Inca Trail

Go to page 217 and complete the questions that relate to the heights climbed and the distances walked along the famous Inca Trail to Machu Picchu.

The Inca road system is often compared to the Roman roads, but the former was built for walking or running, not for wheels. Inca roads were used by the military and for religious reasons, but the main reason for the careful construction was the need to move information from point to point, and to Cuzco in particular. The initial work on the system was completed by the Wari culture before the Inca. Work continued to maintain and extend the system throughout the rule of the Inca.

Many times, the Inca had to cross streams that flowed down the mountains of the Andes. In these situations, they built a grass bridge by weaving *ichu* grass (also called Peruvian feather grass) together to make strong ropes. The ropes were replaced every year by local villagers as part of their tax payments. Each bridge had a manager appointed to ensure the structure was in good condition, as shown in the illustration below.

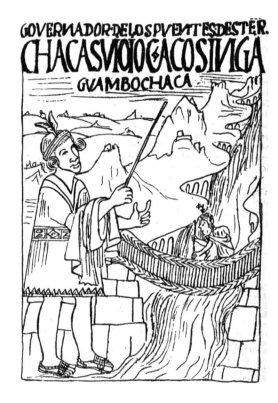

This illustration by Felipe Guáman Poma de Ayala shows the importance and achievements of the construction of bridges. Bridges were necessary across the many streams and chasms that are part of the Andes mountains.

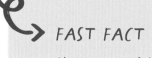
Along some of the more important sections, the Inca placed distance markers. Usually these were every 7 km. These distances were equivalent to one *topo*, the Inca unit of land distance.

A famous section of the Inca trail walked by many people today is a 40 km section that ends at Machu Picchu. It takes four days to walk because much of the road passes through high mountains. Modern walkers need to be very fit like an Inca warrior. Interestingly, this trail is one-thousandth of the total distance of the Inca road system.

Bridges like those built by the Inca are suspended from points on both ends. Today, this type of bridge might have intermediate points making it possible to span a very wide river.

The curve formed by a free-hanging rope, chain, cord or string makes a special curve. The formal mathematical name for this curve is a "catenary". The word *catenary* comes from the Latin word *catena*, meaning a chain. It looks very much like other famous mathematical curves – the parabola and hyperbola. A catenary must be a free-hanging curve without forces pulling it down. So a suspension bridge with cables holding a roadway is not a catenary – the weight of the road changes the catenary to form another mathematical shape called a parabola. The bridge shown here is a true catenary, but we cannot say the Inca invented it. A catenary curve is a naturally occurring entity.

Inca bridges were made of grass woven together to make strong cords. This type of bridge required more ongoing maintenance than the roads. Without knowing it, the Inca were creating free-hanging shapes with some important mathematical properties. The inverted form of these shapes, called catenary arches, are very strong.

ACTIVITY 49:
Creating a catenary

Go to page 218 to explore the catenary curve and discover some of the many places where it can be found. You might be surprised.

Interestingly, the catenary has become a feature of many famous constructions because of its strength when inverted. It is used today by architects to provide strength and, of course, majestic images. One example of the inverted catenary is the Basílica de la Sagrada Família in Barcelona, Spain, designed by Antoni Gaudí. The picture on the opposite page illustrates the amount of open space that can be created using this special mathematical curve. It is a little ironic that the Spanish conquered the Inca civilisation yet the catenary was a feature in Art Nouveau and Modernism, which were prevalent in Spanish architecture during much of the early 20th century. Due to the work of Gaudí and other ingenious architects, the curve can now be found in many places.

An example of an inverted catenary is the famous Basílica de la Sagrada Família in Barcelona, Spain. The building was started in 1882 and is still under construction! The architect was the Catalan architect Antoni Gaudí (1852–1926). There seem to be an endless number of catenary arches in this photograph.

ACTIVITY 50:
Inca civilisation crossword puzzle

Go to page 219 to complete the activity to review this unique culture and the mathematics it used.

Summary

The Inca resided along nearly the full extent of the west coast of South America. They did not have a written language.

Numbers

- Used a base-10 system without symbols
- Showed numbers with knots tied on cords called quipus
- Performed calculations on a *yupana* carved from stone, but the exact use is still a mystery

Geometry

- Were builders of roads and buildings that could withstand powerful earthquakes
- Carved stones to fit perfectly and interlock

Measurement

- Left no evidence of a system or separate units for measurement attributes except for the *topo* (a unit of length)

The Inca

Algebra

- Left no direct evidence of work involving algebra; however, using knots on cords is itself an example of algebra
- May have used algebraic thinking in calculating using the *yupana*

The Inca legacy

The Inca will be remembered for buildings and cities, such as Machu Picchu, in difficult terrain.

Activity 42: *Keeping in contact?*

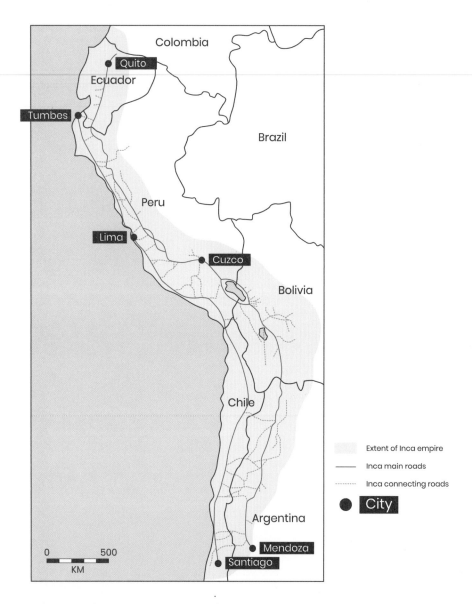

1. What modern-day countries include parts of the Inca empire?

_____   _____

_____   _____

_____   _____

2. Underline the city of Cuzco on the map.

3. Using a ruler, estimate the distance along the main roads from Cuzco to:

    a. Quito, Ecuador – about _____ km.

    b. Mendoza, Argentina – about _____ km.

    c. Tumbes, Peru – about _____ km.

    d. Santiago, Chile – about _____ km.

4. Inca runners (*chasquis*) could relay messages a distance of 320 km in one day. Estimate the number of days it would approximately take to send a message from Cuzco to:

    a. Quito – _____ days.

    b. Mendoza – _____ days.

    c. Tumbes – _____ days.

    d. Santiago – _____ days.

5. If a single runner ran 1,500 m, how many runners would be required to race a distance of 320 km?

                                                         About _____ runners.

6. Estimate the number of runners required to take a message on the 5087 kilometre journey from Quito, Ecuador, to Santiago, Chile.

                                                         About _____ runners.

Activity 43: *Reading a quipu*

1. Write (in words) the three-digit number you see on each cord of this quipu. The value of each digit is greater than zero.

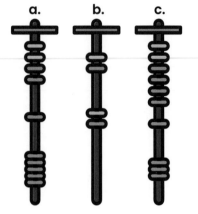

a. _____

b. _____

c. _____

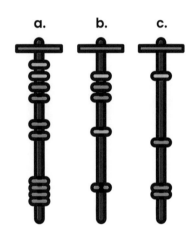

2. Write (in words) the four-digit number you see on each cord of this quipu. Each number has one digit that is zero.

a. _____

b. _____

c. _____

3. Use numerals to write the numbers for each question above in order from least to greatest.

Question 1: _____

Question 2: _____

4. How did you work out the correct order? Did you use words or the pictures?

_____

_____

_____

Activity 44: *Showing a number on a quipu*

1. Draw knots on the quipu to show each of these three-digit numbers.

    a. six hundred and ninety-one

    b. five hundred and seven

    c. five hundred and seventeen

2. Draw knots on the quipu to show each of these four-digit numbers.

    a. two thousand, one hundred and six

    b. three thousand and sixty-one

    c. two thousand, six hundred and four

3. Use numerals to write the numbers for each question above in order from least to greatest.

    Question 1: _____

    Question 2: _____

4. How did you work out the correct order?
Did you use words or the picture?

_____

_____

5. What is the greatest four digit number that can be shown on this quipu with a total of nine knots in each of the thousands, hundreds and tens places? Write the number and draw the quipu picture to match. _____

Activity 45: *Using an Inca "abacus"*

The Inca used a carved stone *yupana* to add numbers. One number was shown by placing stones *on* the *yupana*. Stones for the other number were placed beside and *off* the *yupana*. These stones were then moved onto the *yupana* as shown. Regrouping occurred as required when the stones were moved.

1. Follow the steps to begin adding 437 and 315 on a *yupana*.

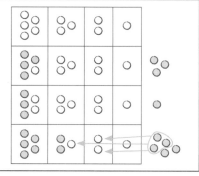

Step 1: Add ones to make a ten **1.** What number is on the *yupana*? _____ **2.** What number is off the *yupana*? _____ **3.** How many ones can be moved onto the *yupana* before regrouping? _____ **4.** How many ones are still off the *yupana*? _____	

Step 2: Regroup ones for ten **5.** Complete this statement for the picture shown here. There are _____ stones on the *yupana* in the first three columns of the row for the ones place. These stones will be traded for _____ stone in the regroup position of the ones row.	

Step 3: Move the ten and left-over ones **6.** After the 3 stones in the ones row have been moved, how many in the ones place will there be on the *yupana*? _____ Go to the next activity page to complete 437 + 315 on an Inca *yupana*.	

Activity 46: *The Inca* yupana *calculating board*

1. Use this *yupana* board and stones or little cubes to help add the following using the Inca method.

| **a.** 437 + 315 = _____ | **b.** 728 + 456 = _____ |

Activity 47: *Analysing angles*

1. Outside Cuzco is a massive structure built with huge stones. The stones are irregular, but do interlock with neatly carved angles. The stones have anywhere from three to 12 sides (and angles). Write the name of the polygon shape that matches the number of angles. Use the labels to help you for some names you may not know.

| nonagon | tetragon | dodecagon | trigon | heptagon |
| hexagon | decagon | hendecagon | octagon | pentagon |

(These labels are all words the Greeks used.)

Number of angles	Name of shape		Number of angles	Name of shape
3			8	
4			9	
5			10	
6			11	
7			12	

2. This special stone found on the Cuzco site has 12 angles.

   a. Sketch the shape below.

   b. Draw a circle around each internal angle you think is acute.

   c. Put an x beside each internal angle you think is obtuse.

Activity 48: *Walking the Inca Trail*

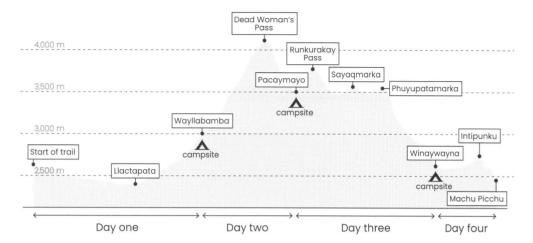

1. The total length of the Inca Trail is about 40 km. Estimate the distance walked on each of the days.

    a. Day 1 is about _____ km.   **c.** Day 3 is about _____ km.

    b. Day 2 is about _____ km.   **d.** Day 4 is about _____ km.

2. Write or sketch the steps you used to make the estimates in Question 1.

3. Find the highest point on the walk. Answer these questions.

    a. On what day is the highest point reached? _____

    b. What is the approximate height of the highest point? _____ m

    c. What is the difference in height between the highest and lowest points on the trail? _____ m

4. Estimate the difference between the heights of each of these pairs of points.

    a. The start of the Inca Trail and campsite at end of Day 1. About _____ m

    b. Campsite Day 1 and campsite Day 2. About _____ m

    c. Campsite Day 2 and campsite Day 3. About _____ m

    d. Campsite Day 3 and Machu Picchu. About _____ m

Activity 49: *Creating a catenary*

1. The catenary is found in nature. What animals made these catenary shapes?

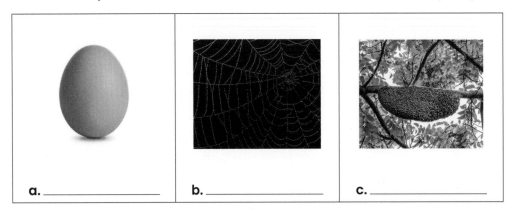

a. _____   b. _____   c. _____

2. Catenary bridges can look different depending on the distance across the gap. What number catenary spans:

    a. the widest gap? _____

    b. the narrowest gap? _____

3. These are inverted catenary curves. Trace over one catenary you can find in each picture.

a. This is found in Arches National Park, USA. The oldest arches found here are in the shape of catenaries.	**b.** This famous arch is located on the bank of the Mississippi river in St Louis, USA.	**c.** Igloos built in far north America are shaped like upside down catenary curves.

Activity 50: *Inca civilisation crossword puzzle*

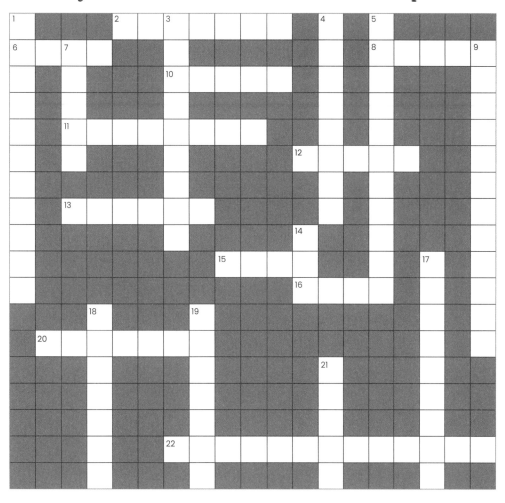

Across

2. Europeans who invaded the new world
6. Indigenous who lived along Pacific coast of South America
8. Mountain range along Pacific coast of South America
10. Material used by Inca to make ropes and bridges
11. Curve made by a hanging chain
12. Rope device used to show numbers
13. The Inca abacus
15. The Inca unit of land distance equal to about 7 km
16. Modern country that was considered the home of the Inca
20. Northernmost country where Inca roads extended
22. Modern name for base-10 numbers (two words)

Down

1. Shape used to show the one in the ones place on a rope (two words)
3. Country with the city of Mendoza
4. Indigenous runner
5. Inca name for "old mountain" (two words)
7. City that was centre of ancient Inca civilisation
9. A continent of the Inca civilisation (two words)
14. The position on a cord that has the greatest place value
17. Linked loops on cords that show more than one in the ones place (two words)
18. The name for a symbol used to write numbers
19. Structures used to cross deep ravines or chasms
21. Country with the city of Santiago

CHAPTER SIX

Early Chinese

When? Where? How?
Who? Why?
I wonder, wonder,
wonder, wonder.

From where did the earliest archaeological evidence of Chinese numerals come?

Why did the Chinese use different coloured rods to represent some numbers?

What is the longest artificial structure ever built?

What is a magic square?
When was the first magic square discovered?

How did a counting board help make Chinese mathematics more efficient?

Early Chinese: *mathematical innovators*

Ancient Chinese mathematics is maybe one of the most overlooked mathematical systems in history. Despite this information gap, there is evidence that mathematics has formed an integral part of Chinese culture since at least the days of the Zhou dynasty (1122 to 256 BCE). At the time, mathematics was considered one

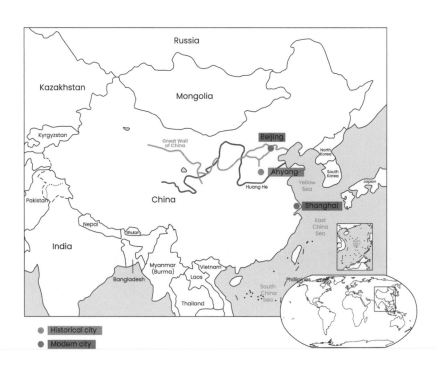

Historical city
Modern city

of the Six Arts that students needed to master to attain the status of a perfect gentleman, akin to the later Western concept of a Renaissance man.

The ancient Chinese were also skilled architects. One of their most impressive feats was constructing the iconic Great Wall of China. This World Heritage–listed site was announced as one of the New 7 Wonders of the World in 2007 along with Chichén Itzá and Macchu Picchu that were discussed in previous chapters. The wall remains substantially intact despite having been built by soldiers and convicts more than 2,000 years ago. Sections of the Great Wall owe their longevity to a rather unusual mortar – glutinous rice flour. As strong and waterproof as cement, this "sticky rice" sealed the bricks so tightly that weeds were never able to grow between them. Construction of the wall began around the seventh century BCE, and sections were still being added up until the time of the Ming dynasty (1368–1644 CE). The wall was originally built to protect China and its Han people in the south from the barbarian nomads in the north. Today, the Great Wall stands as a symbol of national pride and is one of the most popular tourist attractions in the world.

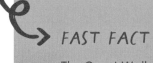

FAST FACT

The Great Wall of China is the longest structure ever built. In fact, the wall is so long that attempts to measure it have only ever resulted in estimates. Walking the full length of the wall can take up to 18 months!

ACTIVITY 51:
Great Wall of China

Go to page 248 and discover the length of the Great Wall of China by completing the estimation activities.

Evolution of Chinese numbers

The earliest-known Chinese numerals came in the form of inscriptions on "oracle bones" more than 3,000 years old from the Shang dynasty, the oldest dynasty in China (also sometimes called the Yin dynasty). It is believed a group of farmers tilling their fields near Anyang (in present-day Henan Province) at the end of the 19th century came across tortoise shells and animal bones with inscriptions on them. These items were later sold to an apothecary, who thought they were the bones of a dragon and endowed with healing properties. Before they could be consumed, the shells and bones were brought to the attention of Chinese scholars who studied the inscriptions they bore. The inscriptions contained information of both the numeral system used back then and the socioeconomic climate of that period.

Tens of thousands of oracle bones – made from animal bones or tortoise plastron (the flat underside of the shell) like this example – have been found and translated. From these nearly 3,000-year-old specimens, scholars have been able to identify some of the first symbols for numerals.

A summary of the symbols that researchers have been able to confidently assign a numerical value. This illustrates a base-10 grouping scheme with different symbols for each size group for the ones, tens, hundreds and thousands.

The number symbols that were identified on these oracle bones eventually evolved into the standard notation shown below.

| 1 | 2 | 3 | 4 | 5 | 6 | 7 | 8 | 9 | 10 |

While these characters were used mainly for record-keeping purposes, they were less suited to computation. The demands of commerce, trade and administration eventually led to the development of a distinctive Chinese place-value number system that involved the use of counting rods. Sticks were arranged in columns from right to left representing increasing powers of 10. Ample evidence confirms the common usage of counting rods from the second century BCE, although they were possibly invented even earlier for fortune-telling practices. The sticks were originally made of bamboo, wood or ivory and were carried in a bag hung at the waist. The length of the sticks, which can be verified by the relics found from this time, ranged from 13.8 to 12.6 cm, gradually shortening as time progressed. The shape also changed with time, from circular (a cylinder shape with a diameter of 0.23 cm) to flat-sided, so that the rods did not roll around as much.

Initially, the rod system consisted of stick symbols from one to nine in vertical alignment. These characters were called *zongs*. Zero was represented by an absence of sticks in the relevant place.

In time, the system became confusing since adequate space was not always left between the symbols to clearly differentiate place values. For instance, the rod numeral "III" could be read various ways if written with different amounts of space between each rod. To overcome this problem, the Chinese developed a second set of rod characters called *hengs*. These characters also represented numbers from one to nine, but the sticks were arranged horizontally. In the modified system, large numbers were shown by alternating between vertical and horizontal characters from one place value to the next. The *zong* characters were used for the ones, hundreds, ten thousands and so on while the *heng* characters represented values in the tens, thousands, hundred thousands and so on places. The *Sunzi Mathematical Manual*, written about 1,500 years ago, stated it more formally:

> Rods that represent the digits of even powers of ten are to be placed vertically. Rods that represent digits of odd powers of ten are placed horizontally.

This removed any confusion around number columns that were written close together.

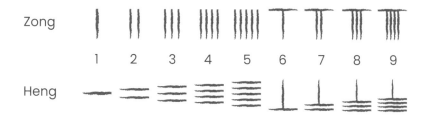

Scholars and merchants adopted the practice of placing rods on a wooden device called a counting board. The board had distinct columns labelled according to their place value, and sticks were manoeuvred around the board to perform calculations. Although the counting board was eventually phased out by the neater and more efficient Chinese abacus (or *suan pan*), some scholars claim that the abacus actually did little to advance Chinese mathematics and may in fact have contributed to its stagnation. Unlike the *suan pan*, counting rods were more than just a counting device; they played an important role in suggesting new approaches to algebraic problems and ways of operating with negative numbers.

Calculating with rods in China began about 2,000 years ago. The rods, used on a board that sat on a table or was placed on the floor, were made from a range of natural resources – wood or bamboo, bone, bronze, or (to be lavish) ivory.

ACTIVITIES 52 & 53:
Working with rod numerals

Go to page 249 and complete the activities to first translate rod numerals to Hindu–Arabic and then draw rod numerals to match familiar numbers.

Negative numbers

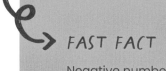

One mathematically interesting feature of the rod number system is the occurrence of red sticks to denote positive numbers and black sticks to denote negative numbers. If coloured rods were unavailable, a negative number was indicated by placing a diagonal rod over the last non-zero digit. There is a strong case that the Chinese are the oldest adopters of negative numbers. Such numbers appeared for the first time in the *Nine Chapters on the Mathematical Art*, which dates from the period of the Chinese Han dynasty (202 BCE to 220 CE).

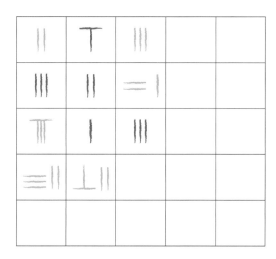

Positive and negative numbers on a counting board.

Although negative rod numerals were not thought of as numbers per se, the Chinese understood how these rods could be used to denote a deficiency in quantity. The Chinese recognition of negative numbers attests to their understanding of how debt operated in commercial trade. Positive numbers were used to indicate how much money was received, while negative numbers indicated the amount of money expended. These figures were then placed on a counting board to work out how much profit or loss there had been over the course of the day.

ACTIVITY 54:
Negative numbers

Go to page 251 to complete the activity to learn how the ancient Chinese represented negative numbers.

Magic squares

Magic squares have been around for many centuries and are well documented throughout Chinese history. At the basic level, magic squares are grids of numbers written in a square array where the sum along each row, column and major diagonal is the same number (called the *magic number*). According to Chinese legend, the first magic square was supposedly marked on the back of a divine tortoise that appeared before Emperor Yu (c. 2200 BCE) when he was standing on the banks of the Yellow River. This 3 × 3 magic square was called Lo Shu, and the numbers 1 to 9 were formed by beads. For the early Chinese, Lo Shu was as fascinating mathematically as it was spiritually. Looking at Lo Shu, pictured below, you can see many interesting numerical patterns, one of which is the symmetrical arrangement of even and odd numbers around the central point.

Lo Shu is thought to be the oldest magic square. It is likely that the legend of it being seen on the back of a tortoise comes from the fact that the ancient Chinese wrote on oracle bones, which were often made from tortoise shells.

16	3	2	13
5	10	11	8
9	6	7	12
4	15	14	1

Through trade with China, magic squares gradually spread to other countries where they took on a range of alternative meanings and depictions. One of the most famous magic squares appears in a 1514 engraving by German Renaissance artist Albrecht Dürer. The piece, shown below, is titled *Melencolia I*, which carries a connotation of sadness. Paradoxically, the artwork is a celebration of mathematical genius. One of the most striking features from a mathematical perspective is the iconic magic square displayed in the top right corner of the picture. The square is magical because the numbers 1 to 16 are written in such a way that the numbers in each row, column and major diagonal add to the same magic number: 34!

The famous *Melencolia I* by Albrecht Dürer is intended to be uplifting although it looks rather like its name – melancholy. It is filled with mathematics involving geometry, measurement and numbers in the 4 × 4 magic square in the upper right. Dürer was one of the main early contributors to this important era when mathematics was first incorporated into art.

Another example of a magic square can be observed on the external facade of the Basílica de la Sagrada Família in Barcelona, Spain. A magic square is shown alongside a sculptural representation of the kiss of Judas, signifying the betrayal of Christ and imminent crucifixion. Interestingly, the square is not magical in the strict mathematical sense, since some numbers have been duplicated to arrive at the magic number of 33 rather than 34. The arrangement of numbers in the square is intentional as 33 is believed to represent Jesus's age when he was crucified.

This square of numbers is on the external west-facing wall of the Sagrada Família cathedral. Although it is not a true magic square, there are some very interesting ways to find the number 33 in the square other than just adding the rows, columns or diagonals.

M
ACTIVITIES 55 & 56:
Magic squares

Go to pages 252 and 253 to explore 3 × 3 squares that may be magic and then investigate the special square on the Sagrada Família cathedral.

While today we are mostly interested in magic squares from a historical perspective, their influence is still felt in our everyday lives. For example, number puzzles are used in schools to teach counting and pattern-recognition techniques, and magic squares overlap with more complicated problem-solving exercises for adults, such as sudoku.

Another proof of the Pythagorean theorem?

We have already seen that several ancient cultures knew about the right-angled triangle and its importance in practical life. Understanding the complexity of these shapes was necessary for handling everyday problems involving geography, engineering and even astrology. There is evidence that the Chinese civilisation had knowledge of Pythagorean triangles, quite possibly before the Greeks did and most probably developed independently.

Much as Euclid did, the Chinese seem to have arranged four identical right-angled triangles around the edge of a square grid so the sides of a large square were equal to the lengths of the hypotenuse of each triangle. The triangles were formed by drawing rectangles at each corner of the grid and then cutting them diagonally in half. For each triangle, the lengths of the two shortest sides were three and four tiles, respectively. The hypotenuse could be discovered by finding out how many tiles made up the large square inside the grid. This could only be done by working out the total area occupied by right-angled triangles and then subtracting this from the total number of tiles in the grid. Once the area of the square was known, it was possible to work backwards to find not only the side length of that square but also (incidentally) the hypotenuse of each right-angled triangle. In the case of the Chinese proof, the length of the hypotenuse (five tiles) could be inferred from the fact that the square occupied an area equal to 25 tiles. There is no direct evidence confirming that this was the method actually used by the Chinese. The only writing that was included with the image was the word "Behold"! The algebra that we associate with the theorem today was not invented until much later.

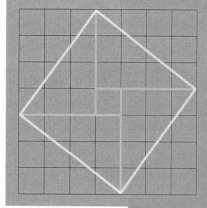

The vertical and horizontal lengths of each triangle in this Chinese right-angled triangle proof are rather easy to find (either three or four). Calculating the length of the diagonal lines – the white lines – was not straightforward for the mathematicians in ancient China. The illustrations suggest that the total number of squares was found by subtracting the four halves of the 3 × 4 rectangles (24) from the total in the picture (49). This gave the number inside the square of white lines and hence the lengths of the sides.

Regardless of when the relationship was first discovered, one fact is certain: the Chinese knew about the basis of the Pythagorean theorem and, furthermore, they knew how to use it. The *Circular Paths of Heaven* treatise explains how to use the theorem to find the depth of pools, or more practically, of rivers and streams. This application influenced Indian mathematicians and subsequently appeared in their writing. We also think that mathematicians in ancient China were able to use the Pythagorean theorem to find the distance to astronomical objects. For example, they could use shadows cast by the sun and angle measurements to estimate the distance from earth to the sun.

ACTIVITY 57:
A special triangle

Go to page 254 to complete the activity.

Algebra – one number with infinity digits

[7] D Blatner (1997) *The Joy of π*, Penguin Books, London, p. 1.

[8] D Wells (1986) *The Penguin dictionary of curious and interesting numbers*, Penguin Books, London, p. 48.

Few symbols have captured the interest of mathematicians as much as π (pi). To some, pi is an utterly frustrating number, "consuming more brainpower, and filling more wastebaskets with discarded theories than any other single number".[7] For others, pi is "the most famous and most remarkable of all numbers".[8] Whichever way you look at it, pi is important for us all to understand. At the simplest level, pi is a value that expresses the ratio of a circle's circumference (the distance around the circle) to its diameter (the distance through the centre of the circle). It is part of a group of numbers called *irrational numbers*, which basically means it cannot be written as a decimal that either terminates or repeats. And yet, we need an approximate value for pi to make calculations about the radius (the distance from the centre of a circle to its edge), circumference and area of a circle – to work accurately and efficiently with the many round shapes and objects in our world.

Early Chinese

Finding pi

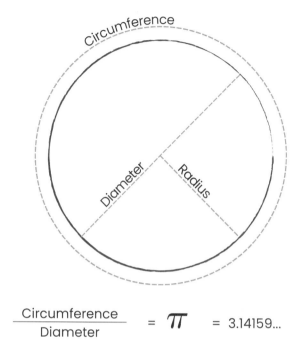

$$\frac{\text{Circumference}}{\text{Diameter}} = \pi = 3.14159...$$

M
ACTIVITY 58:
One number with infinitely many digits

Go to page 255 to complete the activity to learn more about a very special number.

Various attempts have been made throughout history to approximate the value of pi. Even the Bible attempts to do so, with one passage noting the dimensions of a large basin, called Molten Sea, in the Temple of Solomon in Jerusalem, Israel. According to the Old Testament, the basin was "ten cubits from one rim to the other … and a line of thirty cubits did compass it round about" (1 Kings 7:23). This implies a ratio of 30:10 or 3. Many of the earliest approximations of pi came up with a similar number. For example, a Babylonian mathematical tablet found in Susa (in modern-day Iran) in 1936 gives an approximation of pi as 3.125, while a value of 3.1605 appears on a piece of the Rhind Mathematical Papyrus credited to the ancient Egyptians. Both approximations start with 3.1 – impressive estimates for the time but still relatively far off by today's standards. The earliest cultures would have adopted rough estimates of pi mainly by experimentation. This most likely involved stretching a rope between two poles, fixing one in the ground and rotating the other one around it to mark the shape of a circle. The distance around the circle was then compared with the distance through the centre of the circle, producing a ratio that was vaguely framed as three and a little more. While these techniques might have been adequate for primitive engineers and craftspeople, a more accurate calculation was needed in time.

One of the earliest mathematical methods for calculating pi was known as the method of exhaustion. It was first applied by the ancient Greek mathematician Archimedes, who lived more than 2,200 years ago. Calculus had not yet been developed, so Archimedes had to use geometrical approximations for his work. The pictures on the opposite page show how Archimedes calculated the circumference of a circle. He estimated closer and closer to the circumference by drawing regular polygons with an increasing number of sides. The perimeters of these polygons – inside and outside the circle – narrowed down the range of the circumference, and thus pi. As the number of sides increased, the two perimeters would gradually approach the circumference of the circle. Archimedes started with a six-sided hexagon and doubled the number of sides of successive polygons until he had a polygon with 96 sides. This gave him an approximation of 3.14 – the same number that is generally accepted as precise enough when working with pi today.

Early Chinese

Archimedes's exhaustion method

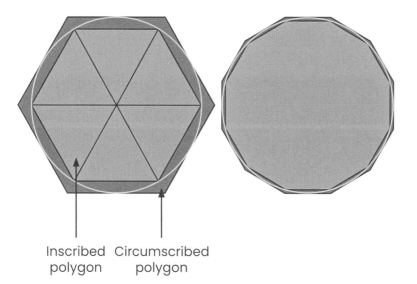

Inscribed polygon   Circumscribed polygon

The ancient Chinese also calculated pi using polygons, but their method was slightly more rigorous. This was partly because the Chinese notation was able to accommodate the basics of trigonometry and finding square and cube roots in a way that the Greeks could not. Around 265 CE, the Chinese mathematician Liu Hui proposed a fast and efficient approximation of pi. The image on the next page summarises Liu Hui's method. Rather than comparing upper and lower bounds, Liu Hui focused only on inscribed polygons. He began by inscribing a hexagon within a circle and breaking the hexagon into distinct triangular portions (marked with dark solid lines).

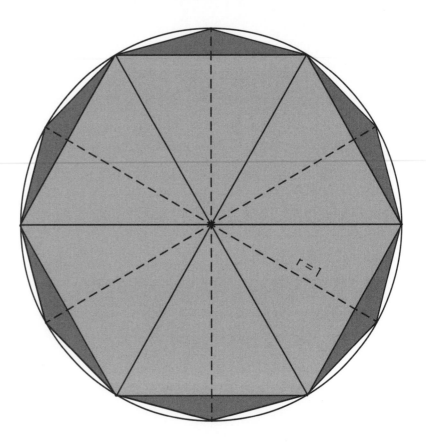

From there, he created polygons with twice as many sides by drawing lines (called bisections) through the middle of each triangle and out to the circle's edge (marked with dotted lines). The dotted lines that occupied the space between the hexagon and the edge of the circle became the edges of a new polygon with twice as many sides: a dodecagon (shaded dark orange). Importantly, the places where the bisecting lines cut through the edges of the hexagon created several right-angled triangles. This allowed certain dimensions, including the perimeter of the newly formed dodecagon, to be calculated using the Pythagorean theorem. Liu Hui continued performing bisections in a similar way until he had created an inscribed polygon with 192 sides. This gave him an approximation of pi of 3.1418 – accurate to three decimal places.

Then, around 200 years later, another Chinese mathematician, Zu Chongzhi, calculated pi to seven digits of accuracy. Adopting Liu Hui's method, he went one step further by calculating the perimeter of inscribed polygons with up to 24,576 sides – it is not difficult to see where this method, usually called Archimedes's *exhaustion* method, gets its name! With this number of sides, Zu Chongzhi determined pi to be about 3.1415929. This was an unprecedented feat, with a more accurate representation of pi not being found until the start of the Middle Ages almost a full millennium later.

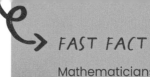

FAST FACT

Mathematicians continue today to compete with one another to find a "better" approximation of pi. This has not changed despite rigorous proofs that have shown the exact value of pi is a decimal that *never ends* and *never starts to repeat*. Newer and more sophisticated technological developments have never falsified those proofs, even after printing numbers with millions, billions and even trillions of decimal places. It is remarkable to think how much time, effort and printing paper has been spent to capture this truly never-ending number! This is the value of pi to 50 places: 3.141592653 5897932384626433833 2795028841971693993 7510.

Number pyramids

Mathematics is governed by patterns. These patterns help to make the most complex mathematical problems easier to work through and solve. Even the earliest cultures recognised this crucial fact. The diagram of Chinese characters shown here is an excellent example of a mathematical pattern. According to George Gheverghese Joseph, it appeared in the book *Si Yuan Yu Jian* (*Precious Mirror of the Four Elements*) by Zhu Shijie, published in 1303 CE.

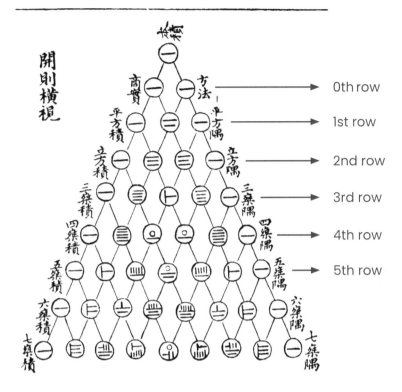

古 法 七 乘 方 圖

開則橫視

0th row
1st row
2nd row
3rd row
4th row
5th row

The Chinese discovered that the triangle was linked to the numbers that are coefficients that preceded each part of a binomial expansion shown below. A binomial is a mathematical expression of the sum of (or difference between) two terms. The coefficients in each of these expanded equations match the numbers in a row of the triangle above. Note that coefficients with a value of one are usually omitted and are not shown here.

$(a + b)^1 = a + b$

$(a + b)^2 = a^2 + 2ab + b^2$

$(a + b)^3 = a^3 + 3a^2b + 3ab^2 + b^3$

$(a + b)^4 = a^4 + 4a^3b + 6a^2b^2 + 4ab^3 + b^4$

Zhu Shijie describes the clear link between the numbers in the triangle and the coefficients, and Joseph provides a complete translation of the explanation in his 1991 book *The Crest of the Peacock: Non-European Roots of Mathematics*. The numbers used by the Chinese enabled them to work with some complex mathematical tasks, such as finding square and cube roots. The process they used for these two tasks involved finding solutions of numerical equations.

A similar number pattern was originally illustrated in a book by Yang Hui published more than 700 years ago. His version was similar, except it used Chinese rod numerals and was drawn out to nine rows. The Chinese appear to have used the number triangle mainly for higher level mathematics involving binomial proofs and expansions.

The triangle, shown here with Hindu–Arabic numerals, is now named after the 17th century mathematician and philosopher Blaise Pascal, who used it to help establish some of the important mathematical principles associated with statistics and probability.

The numbers in Pascal's triangle are generated by counting the number of possible paths to reach each "spot" starting from the apex. As the triangular structure suggests, there are just two possible choices at each decision point. The choice of taking the left or the right branch can be made by tossing a coin with two sides; we usually call the sides heads and tails, but in the example on the next page they are a and b. If the coin is tossed once, two things can happen: a (move to the left) or b (move to the right). If the coin is tossed twice, the possible paths are aa, ab, ba or bb. In this case, there is one way to land on aa, one way to land on bb, and two ways to land on ab. The second row of Pascal's triangle gives these values.

Early Chinese

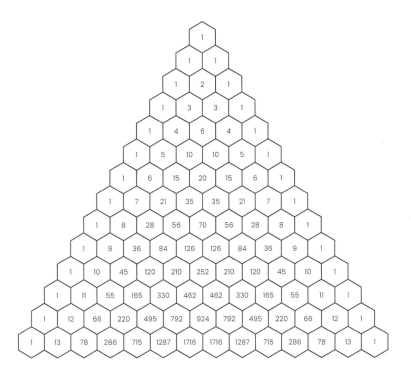

Rows farther down the triangle show how the number of paths to reach each "spot" changes. While the value is always one down the two sides, the numbers in the middle of the rows grow quickly. As a result, it was challenging to work out the numbers every time this information was needed. So it was important to have Pascal's triangle as a ready reference that gave the number of outcomes by row and position in each row. Over time, the range of applications for the numbers found in Pascal's triangle grew and grew, and it has become an invaluable tool.

ACTIVITY 59:
Pascal's triangle

Go to page 256 to investigate the role that this special triangle played in counting combinations of chance occurrences.

ACTIVITY 60:
Chinese mathematics crossword puzzle

Go to page 257 to complete the activity.

Summary

China is one of the most under-represented cultures in the mathematical literature. It is the most populated country in the world and boasts the longest artificial structure ever built.

Numbers

- Used a base-10 system
- Used rod numerals and a counting board for computation
- Adopted negative numbers for commercial purposes like tracking debts
- Created the first 3×3 magic square (known as Lo Shu)

Geometry

- Explored how inscribed polygons could be used to approximate π and the area inside a circle
- Developed a visual proof for the 3:4:5 right-angled triangle as in the *Circular Paths of Heaven* text about 100 BCE

Measurement

- Constructed the Great Wall using advanced formulas for volume, area and proportion
- Used an understanding of right-angled triangles for surveying land

Algebra

- Used basic trigonometry and root extraction techniques to solve for unknown sides of right-angled triangles, which helped them to come up with accurate approximations of π

The Chinese legacy

Chinese mathematicians changed the perception of negative numbers from "absurd" to "useful". Their number pyramids are still used today for improving counting and pattern-recognition techniques.

Activity 51: *Great Wall of China*

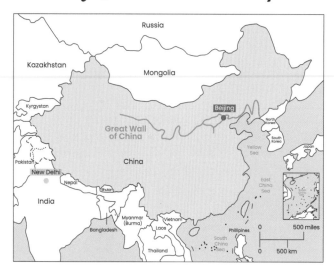

1. Using the above map, estimate the following approximate distances.

a. Beijing to New Delhi is about _____ km.

b. The most eastern point to the most western point of the Chinese mainland is about _____ km.

2. The length of the Great Wall is about 21,196 km. Is it longer or shorter than the distance between Beijing and New Delhi? _____

3. Fill in the blank sections of text below to learn about one of the most amazing structures ever built. Select your answers from the word and number list provided.

The Great Wall of China was built by soldiers and convicts over _____ years ago.

At the time, China had many _____. The Wall was originally built as a divider between the _____ people in the south and the nomads in the north. It is estimated that some _____ labourers died during the wall's construction.

Today, the Great Wall stands as a _____ of national pride. It is the _____ artificial structure ever built, requiring up to 18 months to walk from one end to the other!

Word and number list
symbol
Han
400,000
longest
enemies
2,000

Activity 52: *Reading rod numerals*

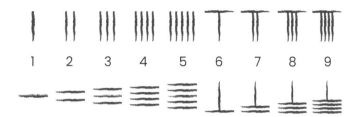

1 2 3 4 5 6 7 8 9

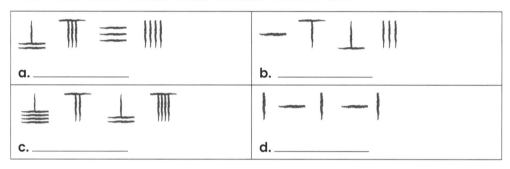

1. Re-write these rod numerals in Hindu–Arabic form.

a. _____	b. _____
c. _____	d. _____

2. Write the Hindu–Arabic numbers for the following rod numerals. Note the empty spaces.

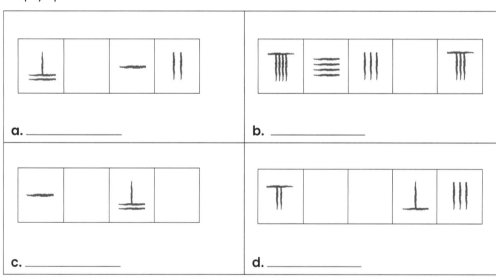

a. _____   b. _____

c. _____   d. _____

3. How many rods are used to show each number in Question 2?

a. _____	b. _____	c. _____	d. _____

Activity 53: *Drawing rod numerals*

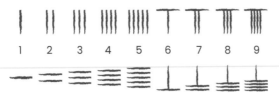

1. Show the following numbers in rod numeral form. Make sure to carefully position your rods in the boxes provided.

a. 234	**b.** 719

c. 5,641	**d.** 8,264

e. 9,090	**f.** 7,000

2. Solve these problems by showing a rod numeral answer. Then write each rod numeral as a Hindu–Arabic number.

a. What is the largest three-digit number you can make with two rods?	**b.** What two 3-digit odd numbers can you make with six rods?

Activity 54: *Negative numbers*

The Chinese are sometimes considered the oldest adopters of negative numbers. To indicate that the number was negative, a rod was placed diagonally over the last non-zero digit.

1. Study the counting rods in each number basket. Using these rods, determine the new position of the basket on the number line. Mark an arrow in the new place.

a.

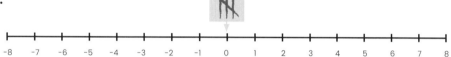

b.

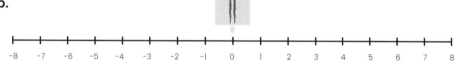

2. Mark an arrow to show the new position of the number basket. Pay attention to the different number intervals and starting positions.

a.

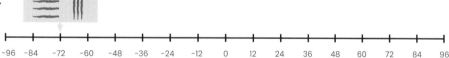

b.

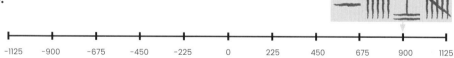

c.

Activity 55: *Magic squares*

The Lo Shu number puzzle is one of the first magic squares ever recorded.

1. Study the Lo Shu magic square below. Write the sum of the numbers along each row, column and major diagonal.

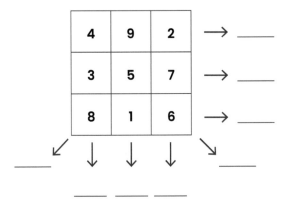

4	9	2	→ _____
3	5	7	→ _____
8	1	6	→ _____

2. Calculate the magic numbers in the following squares.

a.

5	0	7
6	4	2
1	8	3

_____

b.

24	54	12
18	30	42
48	6	36

_____

c.

20	70	60
90	50	10
40	30	80

_____

3. Determine whether the following squares are magic.

a.

34	24	32	10
12	30	22	36
14	28	20	38
40	18	26	16

Magic / Not magic

b.

40	30	20	10
12	34	32	22
24	14	36	26
28	18	16	38

Magic / Not magic

c.

26	16	24	2
4	22	14	28
6	20	12	30
32	10	18	8

Magic / Not magic

Activity 56: *The Sagrada Família number square*

1. The magic number of each square is 33. Use the title of each picture as a clue to find the arrangement that totals 33. All numbers are involved – each solution has an arrangement of four sets of four numbers that add up to 33.

a. Four rows

1	14	14	4
11	7	6	9
8	10	10	5
13	2	3	15

b. Four columns

1	14	14	4
11	7	6	9
8	10	10	5
13	2	3	15

c. Four in each corner

1	14	14	4
11	7	6	9
8	10	10	5
13	2	3	15

d. The diagonals

1	14	14	4
11	7	6	9
8	10	10	5
13	2	3	15

e. Four middle and ends

1	14	14	4
11	7	6	9
8	10	10	5
13	2	3	15

f. Corners of 3 by 3

1	14	14	4
11	7	6	9
8	10	10	5
13	2	3	15

Activity 57: *A special triangle*

Over 2,000 years ago, the Chinese explored some right-angled triangles.

1. Complete the activities in each of these steps. Write the answers or draw on the grid.

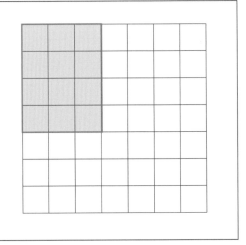

a. What are the dimensions of this large square? _____

b. How many squares are in the total grid? _____

c. How many squares are in the smaller box ▢? _____

d. Draw a diagonal in the ▢ from the top right to the lower left corner.

e. How many squares are in half of the rectangle? _____

2. Three more copies of the rectangle and the first diagonal have been drawn on the grid. Complete these activities.

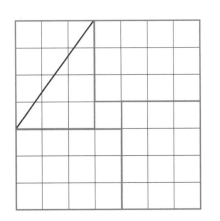

a. Draw one blue diagonal in each ▢. You have now created a blue square.

b. How many grid squares are *outside* the blue square? _____ (Use Question **1e** to help find this answer.)

c. Complete this equation to work out the number of grid squares inside the blue square.

Total in the large square
− number outside the blue square
= number inside the blue square

_____ − _____ = _____

3. If there are 25 grid squares inside the blue square, what is the length of one side of that square? _____

Activity 58: *One number with infinitely many digits*

1. Write the missing word.

a. The distance through the centre of a circle, from edge to edge, is called the _____ of the circle.	**b.** The set of all points equally distant from the centre of a circle is called the _____ of the circle.
c. The distance from the centre of a circle to the edge is called the _____ of the circle.	**d.** If you divide the circumference (C) of a circle by the diameter (d) of the same circle, the value is always the same. This number is the Greek letter _____.

2. You will need a piece of string and a ruler for this activity. Look around. Find five circles. Use the string and ruler to measure the two dimensions. Record the lengths in the table below and calculate their ratio of circumference to diameter.

Object	Circumference (C)	Diameter (d)	$C \div d$

Activity 59: *Pascal's triangle*

1. Imagine you toss a coin that is red (R) on one face and black (B) on the other face. List the possible ways to "get" these outcomes for each number of tosses.

	No Red	1 Red	2 Red	3 Red	4 Red
One toss	B	R	n/a	n/a	n/a
Two tosses	BB	BR, RB	RR	n/a	n/a
Three tosses	BBB	RBB, BRB, BBR			n/a
Four tosses	BBBB				

2. Count the number of outcomes in the table above. Write the numbers in this triangle.

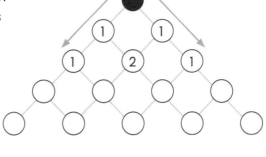

3. What numbers do you think will be written in the next diagonal row?

Activity 60: *Chinese mathematics crossword puzzle*

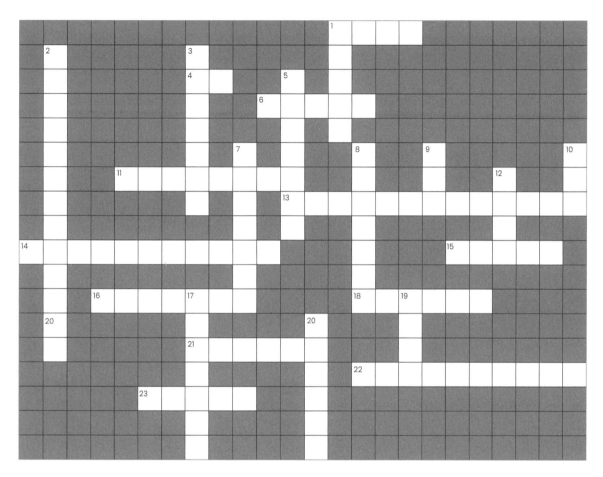

Across

1. The numeral the Chinese did not need
4. Emperor credited with first sighting of a magic square
6. Special name for horizontally aligned rod numerals
11. Divine creature associated with Lo Shu square
13. Wooden device used to perform arithmetic (two words)
14. Earliest evidence of Chinese numeral inscriptions (two words)
15. Name of the 3 × 3 magic square discovered in China (two words)
16. Name given to Chinese abacus (two words)
18. Estimate of pi from a basin in the _____ of Solomon
21. The shape of a group of numbers with a magic value
22. The shape of the triangle used by the Chinese that showed the rule later proved by the Greeks (two words)
23. Colour of rods denoting negative quantities

Down

1. Special name for vertically aligned rod numerals
2. The name given to the distance around the edge of a circle
3. The period between 1122 and 256 BCE was called the Zhou _____
5. If you excelled at the six arts you were a "_____ gentleman"
7. The value of the magic number in the Lo Shu square
8. The Great Wall is the _____ structure ever built
9. Rod numerals were organised in base _____ system
10. Colour of rods denoting positive quantities
12. Name given to sticks used for counting purposes
17. Modern name for the number triangle first created in China
19. Dynasty that added to the Great Wall about 500 years ago
20. The river where the first magic square was discovered

Bibliography

Aczel AD (2015) *Finding zero: a mathematician's odyssey to uncover the origins of numbers*, St Martin's Press, New York.

Anderson G (n.d.) *The Inca yupana and a mathematical society*, accessed 21 March 2022. https://www.arcgis.com/apps/Cascade/index. html?appid=30262eade42b46fd9169c59900cf29e2

Ascher M (1991) *Ethnomathematics: a multicultural view of mathematical ideas*, Brooks Cole Publishing Company, Pacific Grove, California.

Ascher M (2002) *Mathematics elsewhere: an exploration of ideas across cultures*, Princeton University Press, New Jersey.

Ascher M and Ascher R (1981) *Code of the quipu: a study in media, mathematics, and culture*, University of Michigan Press, Ann Arbor.

Aveni A (1989) *Empires of time: calendars, clocks and cultures*, Kodansha International, New York.

Ball J (2017) *Wonders beyond numbers: a brief history of all things mathematical*, Bloomsbury Publishing, London.

Ball WWR (1912) *A short account of the history of mathematics*, Macmillan, London.

Baumgart JK, Deal DE, Vogeli BR and Hallerberg AE (eds) (1989) *Historical topics for the mathematics classroom, National Council of Teachers of Mathematics*, Reston, Virginia.

Bazin M (1995) *Math across cultures*, Exploratorium, San Francisco.

Blatner D (1997) *The joy of π*, Penguin Books, London.

Boyer CB (1968) *A history of mathematics*, John Wiley & Sons, New York.

Brink D (2011) 'Incan and Mayan mathematics', in Greenwald SJ and Thomley JE (eds) *Encyclopedia of mathematics and society*, Salem Press, Hackensack, New Jersey.

Britannica, The Editors of Encyclopaedia (updated 2 February 2022) *Quipu*, Encyclopedia Britannica website, accessed 21 March 2022. https://www.britannica.com/technology/quipu

Burnett J (2005) *Sights, sounds and symbols: classroom activities on the history of numbers*, ORIGO Education, Brisbane.

Burnett J and Irons C (1996) *Egyptian genius*, Mimosa Publications, Melbourne.

Burnett J and Irons C (1998) *Mathematics of the Americas*, Mimosa Publications, Melbourne.

Burton DM (2005) *The history of mathematics: an introduction*, 6th edn, McGraw Hill, New York.

Cartwright M (15 September 2014) *Inca civilization*, World History Encyclopedia website, accessed 21 March 2022. https://www.worldhistory.org/Inca_Civilization/

Cartwright M (8 May 2014) *Quipu*, World History Encyclopedia website, accessed 21 March 2022. https://www.ancient.eu/Quipu/

Catepillán X and Szymanski W (2012) 'Counting and arithmetic of the Inca', *Revista Latinoamericana de Etnomatemática*, 5(2):47–65.

Clawson CC (1994) *The mathematical traveler: exploring the grand history of numbers*, Plenum Press, New York.

Dilke OAW (1987) *Reading the past: mathematics and measurement*, British Museum Press, London.

Dohrn-van Rossum, G (1996) *History of the hour: clocks and modern temporal orders*, University of Chicago Press.

Eagle R (1995) *Exploring mathematics through history*, Cambridge University Press.

Eves H (1969) *An introduction to the history of mathematics*, 3rd edn, Holt, Rinehart, Winston, New York.

Fauvel J and Gray J (eds) (1990) *The history of mathematics: a reader*, Macmillan, London, and Open University, Milton Keynes.

Flegg G (1989) *Numbers through the ages*, Macmillan, London.

FunMaths for High School (updated 15 April 2018) *Make a 6-pointed star mandala*, FunMaths.com, accessed 21 March 2022. http://www.funmaths.com/fun_math_projects/6-pointed_star_madala.htm

Gillings RJ (1972) *Mathematics in the time of pharaohs*, MIT Press, Cambridge, Massachusetts.

Gullberg J (1996) *Mathematics: from the birth of numbers*, WW Norton, New York.

Hewitt PG (2017) 'Focus on physics: the delightful catenary curve', *Science Teacher* 84(4):14–16.

Hodgkin L (2005) *A history of mathematics: from Mesopotamia to modernity*, Oxford University Press.

Hogben L (1960) *Mathematics in the making*, Galahad Books, London.

Hogben L (1993) *Mathematics for the million: how to master the magic of numbers*, 4th edn, WW Norton, New York.

Hogben LT (1968) *The wonderful world of mathematics*, Macdonald, London.

Hollingdale S (1989) *Makers of mathematics*, Penguin Books, London.

Ifrah G (1985) *From one to zero: a universal history of numbers* (Bair L trans), Viking Penguin, New York.

Imhausen A (2016) *Mathematics in ancient Egypt: a contextual history*, Princeton University Press, New Jersey.

Irons C and Burnett J (1994) *Mathematics from many cultures*, Mimosa Publications, Melbourne.

Jackson T (2017) *Numbers: how counting changed the world*, Shelter Harbor Press, New York.

Jorge MC, Williams BJ, Garza-Hume CE and Olvera A (2011) 'Mathematical accuracy of Aztec land surveys assessed from records in the *Codex Vergara*', *PNAS* 108(37):15053–57. https://doi.org/10.1073/pnas.1107737108

Joseph GG (2010) *The crest of the peacock: non-European roots of mathematics*, 3rd edn, Princeton University Press, New Jersey.

Kaplan R (2000) *The nothing that is: a natural history of zero*, Oxford University Press.

Katz VJ (2004) *A history of mathematics: brief version*, Pearson Addison-Wesley, Boston, Massachusetts.

Katz VJ (ed.) (2007) *The mathematics of Egypt, Mesopotamia, China, India, and Islam: a sourcebook*, Princeton University Press, New Jersey.

Kaupp K (2013) 'Math: the age-old question', *Pi in the Sky* 17:8–14.

Kirchhoff P (1943) 'Mesoamérica: sus límites geográficos, composición étnica y caracteres culturales', *Acta Americana* 1:92–107. Cited in Henderson JS (1997) *The world* of the ancient Maya, 2nd edn, Cornell University Press, Ithaca, New York.

Leonard M and Shakiban C (2010) 'The Incan abacus: a curious counting device', Journal of Mathematics and Culture 5(2):81–106.

Locke LL (1923) The ancient quipu or Peruvian knot record, American Museum of Natural History, New York.

MacQuarrie K (30 November 2020) Quipu: the ancient computer of the Inca civilization, Peru for Less website, accessed 21 March 2022. https://www.peruforless.com/blog/quipu/

Mankiewicz R (2000) The story of mathematics, Princeton University Press, New Jersey.

Marianne (5 December 2016) Maths in a minute: the catenary, Plus website, accessed 21 March 2022. https://plus.maths.org/content/maths-minute-catenary

Martzloff J (1997) A history of Chinese mathematics, Springer-Verlag, Berlin.

McLeish J (1992) Number: the history of numbers and how they shape our lives, Fawcett Columbine, New York.

Menninger K (1969) Number words and number symbols: a cultural history of numbers, Dover Publications, New York.

Michalowicz KD and Katz VJ (eds) (2020), Historical modules for the teaching and learning of mathematics (ebook), Mathematical Association of America, Washington DC. https://www.google.com.au/books/edition/Historical_Modules_for_the_Teaching_and/pWXxDwAAQBAJ

Nelson D, Joseph GC and Williams J (1993) Multicultural mathematics: teaching mathematics from a global perspective, Oxford University Press.

Neugebauer O (1957) The exact sciences in antiquity, 2nd edn, Dover Publications, New York.

Norman JM (n.d.) The 'Dresden Codex,' the earliest

surviving book written in the Americas, Jeremy Norman's HistoryofInformation.com, accessed 21 March 2022. https://historyofinformation.com/detail.php?id=1571

O'Connor JJ and Robertson EF (updated November 2000) *Baudhayana*, MacTutor website, accessed 21 March 2022. https://mathshistory.st-andrews.ac.uk/Biographies/Baudhayana/

Reimer D (2014) *Count like an Egyptian: a hands-on introduction to ancient mathematics*, Princeton University Press, New Jersey.

Reimer W and Reimer L (1992) *Historical connections in mathematics: resources for using history of mathematics in the classroom*, vol. I, AIMS Educational Foundation, Fresno, California.

Reimer W and Reimer L (1993) *Historical connections in mathematics: resources for using history of mathematics in the classroom*, vol. II, AIMS Educational Foundation, Fresno, California.

Reimer W and Reimer L (1995) *Historical connections in mathematics: resources for using history of mathematics in the classroom*, vol. III, AIMS Educational Foundation, Fresno, California.

Revell T (14 September 2017) *History of zero pushed back 500 years by ancient Indian text*, New Scientist website, accessed 21 March 2022. https://www.newscientist.com/article/2147450-history-of-zero-pushed-back-500-years-by-ancient-indian-text/

Rohr RRJ (1970) *Sundials: history, theory and practice* (Godin G trans), University of Toronto Press.

Rooney A (2017) *The story of mathematics: from creating the pyramids to exploring infinity*, Arcturus Publishing, London.

Sarmiento de Gamboa, P (2007) *The history of the Incas*

[1572] (Bauer BS and Smith V trans), University of Texas Press, Austin.

Schwartzman S (1994) *The words of mathematics: an etymological dictionary of mathematical terms used in English*, Mathematical Association of America, Washington DC.

Shook EM (1960) 'Tikal Stela 29', Expedition Magazine, 2(2):28–35.

Struik DJ (1954) *A concise history of mathematics*, G Bell & Sons, London.

Studious Guy (n.d.) *10 real-life examples of triangle*, Studious Guy website, accessed 21 March 2022. https://studiousguy. com/10-real-life-examples-of-triangle/

Swetz FJ (2017) 'Mathematical treasure: land surveys in the *Codex Vergara*', *Convergence* June 2017, accessed 21 March 2022. https://www.maa.org/press/periodicals/ convergence/mathematical-treasure-land-surveys-in-the-codex-vergara

Swetz FJ (2018) 'Mathematical treasure: the precious mirror of Zhu Shijie', *Convergence* June 2018, accessed 21 March 2022. https://www.maa.org/press/periodicals/convergence/ mathematical-treasure-the-precious-mirror-of-zhu-shijie

Swetz FJ (ed.) (1994) *From five fingers to infinity: a journey through the history of mathematics*, Open Court, Chicago, Illinois.

Swetz FJ and Kao TI (1977) *Was Pythagoras Chinese? An examination of the right triangle theory in ancient China*, Pennsylvania State University Press, University Park.

Thompson JES (1954) *The rise and fall of Maya civilization*, University of Oklahoma Press, Norman.

Tun M (9 September 2014) 'Yupana', in Selin H (ed.) *Encyclopaedia of the history of science, technology, and medicine in non-Western cultures*, Springer, Dordrecht. https://doi.org/10.1007/978-94-007-3934-5_10273-1

University of Oxford (14 September 2017) *Earliest recorded use of zero is centuries older than first thought*, University of Oxford website, accessed 21 March 2022. https://www.ox.ac.uk/news/2017-09-14-earliest-recorded-use-zero-centuries-older-first-thought

Vedic Knowledge Online (n.d.) *Sanskrit numbers*, Veda.Wikidot.com, accessed 24 March 2022. http://veda.wikidot.com/sanskrit-numbers/comments/show

Wells D (1986) *The Penguin dictionary of curious and interesting numbers*, Penguin Books, London.

Yadav N and Vahia M (2011) *Indus script: a study of its sign design*, Harappa.com, accessed 24 March 2022. https://www.harappa.com/content/indus-script-study-its-sign-design

Yan L and Shiran D (1987) *Chinese mathematics: a concise history*, Clarendon Press, Oxford

Ye X (12 March 2016) *The search for the value of pi*, The Conversation website, accessed 21 March 2022. https://theconversation.com/the-search-for-the-value-of-pi-55744

Yong LL and Tian Se A (2004) *Fleeting footsteps: tracing the conception of arithmetic and algebra in ancient China*, World Scientific, Singapore.

Zaslavsky C (1996) *The multicultural math classroom: bringing in the world*, Heinemann, Portsmouth, New Hampshire.

Image Credits

Page 10
— Science Photo Library / Alamy Stock Photo

Page 15 (clockwise from middle)
— Reinhard Dirscherl / Alamy Stock Photo
— "Seven Shades of Purple" by Daina Taimina (Cornell University, Ithaca, NY), photo © Daina Taimina
— Natalie-claude / iStock
— Pixelfit / iStock

Page 17
— AlexLMX / iStock

Page 18
— The Picture Art Collection / Alamy Stock Photo

Page 28
— Ionel Sorin Furcoi / Alamy Stock Photo

Page 29
— Dorling Kindersley Ltd / Alamy Stock Photo

Page 31
— www.BibleLandPictures.com / Alamy Stock Photo. Photographer: Zed Radovan

Page 39
— iStock/com/Species125

Page 45
— agefotostock / Alamy Stock Photo

Page 49
— BC. 021355. [YBC 7290] Courtesy of the Yale Peabody Museum, Babylonian Collection. Photography by Klaus Wagensonner

Page 202
— Earleliason / iStock

Page 203
— DrMonochrome / iStock

Page 204
— Stephen Wood / Alamy Stock Photo

Page 205 (clockwise from middle)
— The Picture Art Collection / Alamy Stock Photo
— Science Photo Library / Alamy Stock Photo

Page 206
— Westend61 GmbH / Alamy Stock Photo

Page 207
— robertharding / Alamy Stock Photo

Page 216
— DrMonochrome / iStock

Page 218 (from left)
— Nastco / iStock
— Fotogaby / iStock
— winlyrung / iStock
— Paola Giannoni / iStock
— DenisTangneyJr / iStock
— vovashevchuk / iStock

Page 223
— Salimbangla / iStock

Page 225
— Istock.com / kool99

Page 228
— The Picture Art Collection / Alamy Stock Photo

Page 232
— FineArt / Alamy Stock Photo

Page 233
— Mira / Alamy Stock Photo

Page 243
— Art Collection 3 / Alamy Stock Photo

Acknowledgements

The production of a book for a new entity is a mammoth task. In this case, two groups of individuals have contributed heavily to this first publication of the Mathema Publishing Trust. The first are the individuals at the Trust who have been involved with preparation and initial presentation of the manuscript. The second is the team at Hardie Grant who took the writing and initial design work and shaped these elements into an exciting and informative book.

At Mathema Publishing Trust, we must first thank Rosemary Irons for the many hours spent reading and reacting from the outset in 2020 – before there was a Trust – until the day the final checking was completed before printing. She provided invaluable feedback so mathematics, the often-maligned subject, could be presented here in a way that can be better understood by the general reader and young learners. We must also thank Chloe Honnef for the editorial suggestions for the overall look and structure of the book. Her questions and comments helped shape the final structure and presentation, particularly the interweaving of descriptive text and supporting activities.

Above all, we must thank Louis Devereaux who worked throughout all stages of the project from the initial look of the book, to selecting an appropriate publishing partner, and then liaising throughout until the end to ensure the book has the look and educational feel that is warranted. Louis excelled in his enthusiastic dedication, friendly tenacity and focused attention to detail. We thank him.

Hardie Grant has established itself as one of the (if not the) most respected publishing houses in Australia. The process for producing the book was clear, and both Courtney Nicholls and Hannah Louey were always available to give us suggestions and support modifications that we wanted to make (even at the last moment). The academic editing provided by Margie Beilharz was superb. She ensured all facts were verified and questioned where necessary. We also thank George Saad for his diligence and patience for some of the design challenges that arose as the book progressed.

Calvin and Robert

About the Authors

About the Authors

Calvin Iron's interest in mathematics began as he was growing up on a small farm in rural Iowa, USA. As a kid driving a tractor, Calvin remembers using mathematics to quickly work out the number of haybales stacked on wagons or posts needed to enclose fields. His formal study of the subject began in 1961, and since then Calvin has dedicated his life to writing, teaching and promoting the good learning techniques that make mathematics interesting and relevant to our everyday lives and world around us.

In the US, Calvin completed multiple mathematics degrees, and while studying for his PhD, he wrote mathematics materials for the National Science Foundation. In 1975 Calvin moved to Brisbane, Australia, to accept a lectureship at the Queensland University of Technology, which he served for nearly 40 years. Cal taught tens of thousands of students and gained a reputation as an outstanding teacher and speaker. His passion for the good teaching of mathematics saw him travel across Australia and the world to host thousands of professional learning sessions for teachers.

Calvin has previously written material for Rigby Education, Mimosa Publications and McGraw-Hill Education. In 1995 he co-founded ORIGO Education, which now has offices in four countries, and for whom he continues to write innovative material. Calvin has also established the Mathema Publishing Trust with the co-author of this book, Robert Natanek.

Robert Natanek was born in Brisbane in 1996 and recently completed an arts/law degree at the University of Queensland. During this time, he received well-deserved recognition for his writing and research skills.

At the completion of his studies, it was only natural that Robert felt an itch to put those writing skills to good use. Although mathematics was not the expected outlet for those skills, Robert's intellectual curiosity was nonetheless piqued by the realisation that mathematics permeates all parts of culture and life in general. This project was also facilitated in large part by Robert and Calvin's mutual passion for travel. In just 25 short years, Robert has travelled to more than 25 different countries and experienced next to the same number of illuminating cultural perspectives. These experiences, along with Robert's writing skills, were integral in bringing this book to life.

Index

Working space

Working space

Working space

Working space

Published in 2022 by Hardie Grant Books an imprint of
Hardie Grant Publishing

Hardie Grant Books (Melbourne)
Ground Floor, Building 1, 658 Church Street
Richmond VIC 3121, Australia

Hardie Grant Books (London)
5th and 6th Floors, 52–54 Southwark Street
London SE1 1UN, United Kingdom

www.hardiegrant.com.au

Hardie Grant acknowledges the Traditional Owners of the Country
on which we work, the Wurundjeri People of the Kulin Nation and the
Gadigal People of the Eora Nation, and recognises their continuing
connection to the land, waters and culture. We pay our respects to
their Elders past and present.

A catalogue record of this book is available from the National Library
of Australia.

The Amazing Beginnings of Mathematics
ISBN 9781743799482

Publication commissioned by Courtney Nicholls
Publication managed by Hannah Louey
Written by Robert Natanek BA, LLB and Calvin J. Irons BA, MA, PhD
Contributions by Chloe Honnef and Rosemary Irons
Edited by Margie Beilharz
Designed by George Saad
Pre-production by Louis Devereaux
Printed in China by Leo Paper Group

Answers to the activities
in this book can be
accessed via the QR
code or by visiting
mathemafoundation.com/
activity-answers